Knowledge Processing for Structural Design

Topics in Engineering

This series provides a rapid and informal dissemination of significant new work in engineering. It is aimed at high level coverage across a broad field of engineering including mechanical, civil, hydraulic and structural, as well as other associated topics. Books in the Series are produced inexpensively in camera-ready form to allow as wide a readership as possible among the engineering community. Typical book size is between 100 and 500 pages.

Topics in Engineering

Volume 25

Edited by C.A. Brebbia and J.J. Connor

Knowledge Processing for Structural Design

B. Kumar

Computational Mechanics Publications
Southampton UK and Boston USA

Series Editors
C.A. Brebbia and J.J. Connor

Author
B. Kumar
Department of Civil Engineering
University of Strathclyde
John Anderson Building
107 Rottenrow
Glasgow G4 0NG
UK

Published by

Computational Mechanics Publications
Ashurst Lodge, Ashurst, Southampton, SO40 7AA, UK
Tel: 44 (0)1703 293223 Fax: 44 (0)1703 292853
Email: CMI@uk.ac.rl.ib
Intl Email: CMI@ib.rl.ac.uk

For USA, Canada and Mexico

Computational Mechanics Inc
25 Bridge Street, Billerica, MA 01821, USA
Tel: 508 667 5841 Fax: 508 667 7582
Email: CMINA@netcom.com.

British Library Cataloguing-in-Publication Data

A Catalogue record for this book is available
from the British Library

ISBN 1-85312-376-5 Computational Mechanics Publications, Southampton
ISBN 1-56252-300-7 Computational Mechanics Publications, Boston
ISSN 0952-5300 Series

Library of Congress Catalog Card Number 94-69715

Printed in Great Britain by Hobbs The Printers, Southampton

Table of Contents

Preface

This book is an investigation into the potential of applying knowledge-based techniques to the automation of the structural design process. Knowledge-based techniques involve incorporating domain-dependent knowledge in a computer program and manipulating it by a separate set of rules often called control rules using some problem-solving techniques from Artificial Intelligence (AI).

The use of AI techniques is demonstrated by developing simple prototypes to assist in the different stages of the structural design process. First of all, a conceptual model for integrated structural design is presented which suggests a number of enhancements to an existing model proposed by earlier researchers. Subsequently, three components of the model are developed using an expert system development tool and incorporating AI problem-solving techniques.

The prototypes developed in this study incorporate knowledge for the design of industrial buildings. The preliminary design module, ALTSEL, contains knowledge for the selection of alternative structural systems for an industrial building. Most of the knowledge contained in this module is obtained by interviewing some practicing engineers. The knowledge elicitation process for ALTSEL resulted in some important conclusions. The other two prototypes developed are called DETDEX and DESCON and are concerned with the detailed design and design reviewing for portal frames respectively. The purpose of this project was not to develop a fully-working system but investigate and illustrate the utility of AI tools and techniques for computer-aided design of structures.

The computer programming in the project mainly involved encoding rules in the Edinburgh PROLOG Blackboard Shell syntax and procedural clauses in Prolog. The structural analysis programs used were written in FORTRAN77. Sample runs of the prototypes running separately as well as running together as an integrated system are also included. All the rules from the prototypes are not given but some of the representative rules from all the prototypes are included in the respective chapters. Also included is a description of the interface between FORTRAN and PROLOG developed in this work.

Acknowledgments

I would like to thank Prof. BHV Topping of Heriot-Watt University for introducing me to the fascinating field of knowledge-based systems; his helpful comments are gratefully acknowledged. I would also like to thank the Edinburgh University General Council for financial assistance during my research.

I am deeply indebted to Dr Paul Chung and Mr. Robert Rae of Artificial Intelligence Applications Institute, Edinburgh University, for their constructive comments and valuable discussions. I would like to thank my research student, Mr. Benny Raphael, for the cover design.

Finally, I want to express my thanks to my family, and my wife Sangeeta, for being wherever and whenever I needed them.

List of Abbreviations

Following is a list of the abbreviations used in the thesis:

AI	Artificial Intelligence
CAD	Computer-aided design
DDB	Dependency-directed backtracking
DE	Domain expert
EPBS	Edinburgh Prolog Blackboard Shell
KBES	Knowledge-based Expert Systems
KE	Knowledge Engineer
KM	Knowledge module
KS	Knowledge source
TMS	Truth Maintenance System

Chapter 1
Introduction

1.1 General Remarks

As a result of years of research in Artificial Intelligence, knowledge-based expert systems (KBESs) have emerged as a most promising application. Areas of early applications of knowledge-based expert systems technology include medical diagnosis, mineral exploration and chemical spectroscopy. Civil Engineering and, indeed, other engineering disciplines were generally slow to respond and it was not until the mid-1980s that a substantial application of knowledge-based expert systems to Civil Engineering emerged in the shape of an expert system for the preliminary design of high-rise buildings called HI-RISE (1). The research community's response to HI-RISE was generally positive. This work reports one of many such studies undertaken immediately after the development of HI-RISE to investigate the potentials of applying this emerging technology to structural design.

1.2 Knowledge-Based Expert Systems

Until the 1960s, the main use of computers was confined to the number-crunching of large volumes of numerical data. KBESs were devised to widen the scope of computing by incorporating domain knowledge in computer programs. Of course, this new breed of programs was devised after realising the limitations of domain-independent problem-solving systems like the General Problem Solver, GPS (2). The main concept behind the KBESs is to separate the domain-dependent knowledge and the domain-independent control rules to manipulate that knowledge. KBESs will be discussed in detail in Section 2.6.1. KBESs are considered suitable for solving problems that require considerable experience, judgement or rules of thumb. Such problems do not generally have an algorithmic solution and are often ill-structured. Design is one such problem.

1.3 Outline of Problem

Although HI-RISE and its later development ALL-RISE (4) have demonstrated the application of knowledge-based techniques to structural design, the real potential of knowledge-based techniques in solving practical problems remains largely questioned. The industry still remains sceptical and unconvinced. Clearly there is a need for further research and development to prove if knowledge-based technology has something worthwhile to offer. The problems that HI-RISE can solve are seen to be trivial and the solutions generated by it mostly obvious (3). But, HI-RISE can certainly be seen to be an important first step. Obviously, the effort is needed in developing more sophisticated systems that can solve more difficult problems.

This book is concerned with primarily the following two things:

1. to identify the types of problems encountered in structural design; and
2. to identify the role that knowledge-based systems technology can play in automating the solution of some of those problems.

Since knowledge-based systems utilise problem-solving methods from Artificial Intelligence one important goal of this work is to investigate their utility in developing KBESs for structural design. The particular methods to be investigated are dependent on the stages in the design process and the nature of problem-solving in those stages.

1.4 Aims and Objectives

The structural design process usually consists of preliminary design, structural analysis and detailed design. In order to be able to develop knowledge-based expert systems for structural design, the whole process has to be first modelled in appropriate terms. One such model, DESTINY, was proposed by Sriram (4,5). The same model will be examined to identify its limitations and develop it further. Subsequently, different components of the model will be developed. HI-RISE is taken as a starting point and a similar prototype for preliminary design will be developed using other representation formalism and problem-solving methods. In the detailed design stage, most of the work has concentrated upon knowledge-based standards processing. In this area, the aim is to identify the limitations of the existing work and suggest improvements.

One important feature of any design is the feedbacks from one design stage to another. In the final part of this work, a knowledge-based prototype to assist in this aspect of design is to be developed. These prototypes are not intended to be fully-working systems but only to illustrate the applications of some of the techniques from AI for the automation of the structural design process. The prototypes will be developed to identify the AI techniques that may be required to solve problems in the different stages of design. The example domain of the prototypes will be the design of industrial buildings. The preliminary design module suggests alternative structural systems for an industrial building while the other two modules deal with the design of portal frames. The detailed design module will include provisions from BS5950 for the design of portal frames.

The main objective of this study is to ascertain the potential for applying some tools and techniques from Artificial Intelligence for automating the solution of problems encountered in structural design that may not have a purely algorithmic solution and may require non-numerical manipulation of a knowledge-base.

1.5 Layout of Work

The remainder of this work is organised in seven chapters and eight appendices as described below:

Chapter 2 - This chapter describes some of the important problem-solving methods from Artificial Intelligence. The emergence of knowledge-based expert systems as a powerful problem-solving technique is also discussed in some detail. Some of the popular tools for the development of expert systems are also described briefly. The particular tool, Edinburgh Prolog Blackboard Shell, used in this work is described in some detail.

Chapter 3 - This chapter reviews some of the works of knowledge-based expert systems in Structural Engineering. Some of the earlier pioneering work on expert systems in other domains is also included. During the course of this research, there have been so many papers and reports published that it becomes practically impossible to review all of them and only the ones that have some relevance to this study are considered.

Chapter 4 - This chapter presents a conceptual model for integrated structural design called INDEX. The model is based on an earlier model called DESTINY and suggests some enhancements to it.

Chapter 5 - This chapter describes the development of the preliminary design module of INDEX called ALTSEL. The use of the expert system shell used to develop this prototype is also discussed. Some lessons learnt from the knowledge elicitation process for the development of this module are also discussed.

Chapter 6 - This chapter contains a description of the detailed design module of INDEX called DETDEX. The main thrust of this chapter is to present a representation scheme for the codes of practice for engineering design and illustrate its use by the development of a generic processor. The representation scheme suggested is thought to be general and the same processor may be used to process any code of practice represented as suggested.

Chapter 7 - This chapter describes the design reviewing module of INDEX called DESCON. The first part of the chapter presents an examination of a discussion between practicing engineers to identify the problems encountered in practice and also ways of solving them. The conclusions drawn from this part form the basis for the development of DESCON.

Chapter 8 - This chapter identifies the conclusions drawn from the whole study. The conclusions drawn from the development of each prototype are listed first followed by some general conclusions from the whole work. Some suggestions for future research are also given.

Appendix I - This appendix contains a description of the interface between Prolog and Fortran developed in this work to interact with structural analysis programs.

Appendix II - This appendix contains a sample run of a rule-based standards processor developed in the earlier part of this work.

Appendix III - This appendix contains a sample run of the preliminary design prototype, ALTSEL.

Appendix IV - This appendix contains a listing of the conflict-resolution for firing the rules.

Appendix V - This appendix contains a sample run of the standards processor prototype.

Appendix VI - This appendix is a run of the design reviewing prototype on its own.

Appendix VII - This appendix explains a sample run of all the prototypes running as an integrated system illustrating the feedbacks in design.

Appendix VIII - This appendix is a listing of a run of a Fortran program developed in the earlier part of the work for the design of plate girders. The program exhibits some crude learning behaviour by utilising feedbacks from past designs.

References

1. **Maher, M.L. and Fenves, S.J.**, *"HI-RISE: An Expert System for the Preliminary Design of High Rise Buildings"*, Report No. R-85-146, Department of Civil Engineering, Carnegie-Mellon University, Pittsburgh, U.S.A., 1985.

2. **Rich, Elaine**, *"Artificial Intelligence"*, McGraw Hill Book Company, 1986.

3. **Maher, M.L.**, *HI-RISE and beyond: Directions for Expert Systems in Design,,* Computer-aided Design, Volume 17, Number 9, London, pp 420-426, 1985.

4. **Sriram, D.**, *"Knowledge-based Approaches to Structural Design"*, Ph.D. Dissertation, Department of Civil Engineering, Carnegie-Mellon University, Pittsburgh, U.S.A., 1986.

5. **Sriram, D.**, *"Knowledge-based Approaches for Structural Design"*, Computational Mechanics Publications, Southampton and Boston, 1987.

Chapter 2

Artificial Intelligence Tools and Techniques

2.1 Introduction

One of the most important research contributions to Artificial Intelligence (AI) has been the development of effective problem-solving methods for relatively *hard* problems. Another important contribution has been the development of *tools* for the development of knowledge-based systems such as expert system shells. The main objective of this chapter is to give some idea about the different terminologies and also the theoretical aspects of AI touched upon in the later chapters. The descriptions are brief as there are scores of authoritative books (2-6,12) available for detailed treatment of the subject. AI is such a vast and new field that the researchers have yet to agree to even a universal definition of the subject. Some definitions of AI will be given to highlight the degree of dissent and disagreement in the field. One section describes the Truth Maintenance System of Doyle (1) which supports non-monotonic reasoning. The description of the TMS is included because non-monotonic reasoning plays an important role in some parts of this work as discussed in Chapters 4 and 7. Brief descriptions of some of the popular tools for building knowledge-based expert systems will also be given. The particular expert system shell used in this research, viz. the Edinburgh Prolog Blackboard Shell, will be described in some detail to enable the appreciation of its use described in later chapters.

2.2 Artificial Intelligence and Related Techniques

2.2.1 Definition of Artificial Intelligence

It is difficult to find a universal definition of AI. The literature is full of different (and often contradicting) definitions of AI. Some consider it as a science of building thinking machines. Nilsson (2) defines AI as:

> *"a science whose goal is to build machines that perform tasks normally requiring human intelligence."*

On the other hand, some others consider it as being the science and arts of building intelligent software. Boden (3) defines AI as:

> *"the study of the use of programs and programming techniques to cast light on the principles of intelligence in general and human thought in particular."*

Winston (4) defines AI as:

> *"the study of ideas which enable computers to do the things that make people seem intelligent."*

A more general definition was provided by Charniak and McDermott (5):

> *"AI is the study of mental faculties through the use of computational models."*

Rich (6) suggests another definition:

> *"AI is the study of how to make computers do things at which, at the moment, people are better."*

Thus there are as many definitions of AI as there are books about the subject. There seems, however, to be a certain degree of unanimity in all of them. Researchers generally consider AI as the art or science of building intelligent machines either by developing intelligent software or hardware or both. It is beyond the scope of this research to provide a new definition of AI or to present a philosophical criticism of the existing research. One point, however, that should be obvious is that this project is concerned with the development of *intelligent* software (not hardware) using the existing AI tools and techniques, particularly *knowledge-based system technology.*

2.2.2 AI Techniques

Having given some definitions of AI, a brief discussion of what is or should be an AI technique follows.

An AI technique may be considered to be a method that exploits *knowledge* that should be represented in such a way that (6) it may:

1. capture generalisations;
2. be understood by people who provide it;
3. be easily modified;
4. be used in a great number of ways; and
5. be used to help overcome problems regarding *combinatorial explosion* by narrowing down the range of possibilities to be considered.

Some of the important AI techniques are (6):

1. Search - which provides a way of solving problems for which no more direct approach is available and in addition a framework into which any direct technique that is available can be embedded.
2. Use of knowledge - which provides a way of solving complex problems by exploiting the structure of the objects that are involved.
3. Abstraction - which provides a way of separating important features and variations from the many unimportant ones that will otherwise overwhelm any process.

Out of these three, this work is mainly concerned with the use of knowledge. However, some use of the other two techniques of *search* and *abstraction* is made within the overall framework as well. Within the area of *use of knowledge*, the following three sub-areas may be identified:

1. knowledge representation;

2. knowledge manipulation; and

3. knowledge acquisition.

Again, out of these three sub-areas this work is mainly centered on the second one, ie., *knowledge manipulation*. In addition, however, *knowledge acquisition* also plays an important role in the development of one module, ALTSEL, of the prototype knowledge-based system developed as part of this work.

2.3 Problem-Solving Methods in AI

2.3.1 Introduction

Much of the research in AI has concentrated on problem-solving. Most problem-solving methods use the notion of trial and error *search*. That is, these methods solve the problem by searching for a solution in a *space* of possible solutions. The key terms to note in this approach are *search* and *space*. Another approach decomposes a problem into a number of subproblems. The following sections present brief descriptions of these two alternative approaches.

2.3.2 State-space method

The 15-puzzle will be considered as an example to explain the notion of problem-solving as a trial and error search. Although the 15-puzzle does not represent more complex problems such as design, it does, however, give some insight into solving problems that require *intelligence*. To quote Minsky (7), "it is not that the games and mathematical problems are chosen because they are clear and simple; rather they give us, for the smallest initial structures, the greatest complexity so that one can engage some really formidable situations after a relatively minimal diversion into programming." Research in solving problems and games has generated and refined many problem solving ideas that are also genuinely useful on less frivolous tasks.

One of the simplest approaches to finding a solution for the 15-puzzle would be to try various moves until the goal configuration is found. This approach again involves trial and error search. Starting with the initial configuration, applicable moves constantly keep transforming the configuration until the goal configuration is reached. It is useful to introduce the notions of *states* and *operators* at this stage. For this particular example, a problem state is any particular configuration. The initial and final configurations are called the initial and goal states respectively. Between the goal state and the initial state there are many intermediate states. In 15-puzzle, an *operator* is any applicable move. An operator is basically used to transform one state into another. Any problem-solving method that uses the notions of states, search and operators is said to use a state-space approach.

2.3.3 Problem-reduction methods

In this approach, a problem is broken into a number of subproblems. Design is of this class of problem. This approach is applicable to relatively more complex tasks which are too complex and large to be tackled at once and so the problem is decomposed into a number of subproblems by representing it as an AND/OR graph. An AND/OR graph is generated by the application of problem-reduction operators to different problem states. In this approach, the solutions to the different subproblems form the solution to the problem. Each of the subproblems are individually solved using the state-space or possibly the problem-reduction method. This approach of breaking a problem into sub-problems and then solving each one of them separately is called the problem-reduction approach.

2.3.4 Search Methods

As discussed in Section 2.3.1, search plays a major role in most problem-solving methods in AI. The following sections present a brief description of the important search methods.

2.3.4.1 Breadth-first Search

To explain breadth-first search, the trees shown in Figure 2.1(a,b and c) will be considered. The trees depict the different states of a problem starting from the initial state and ending at the goal state. The intermediate states could be generated using

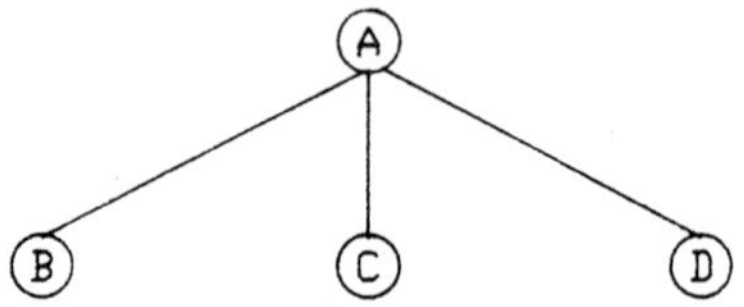

Figure 2.1 (a)

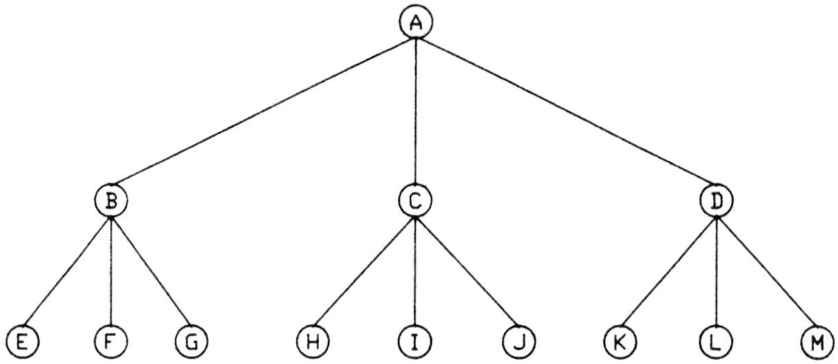

Figure 2.1 (b)

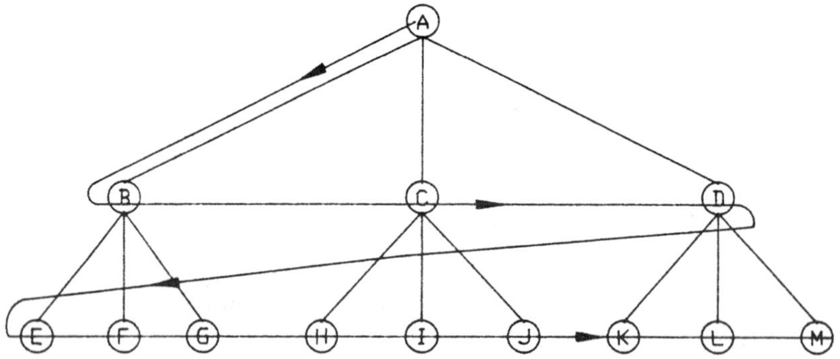

Figure 2.1 (c)

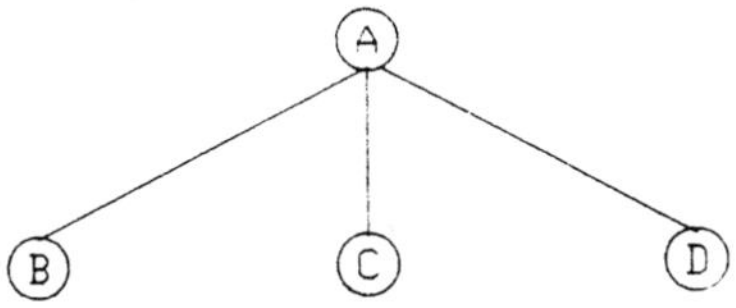

Figure 2.2 (a)

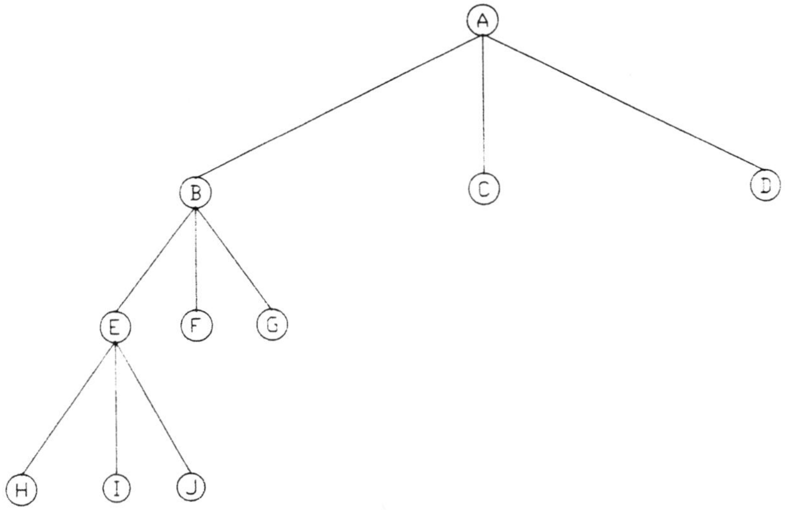

Figure 2.2 (b)

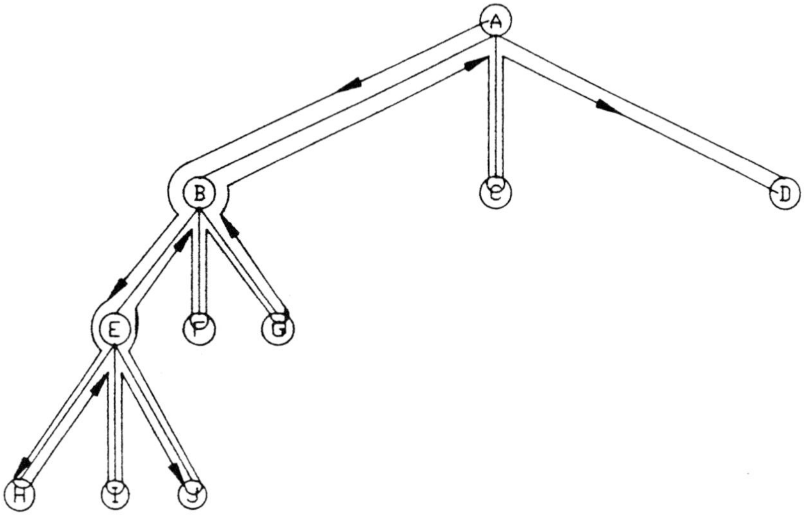

Figure 2.2 (c)

various approaches. One approach is to apply the applicable rules to the initial state (node A) to generate the intermediate state shown in Figure 2.1(a). Again, applying the applicable rules to these nodes will generate more nodes as shown in Figure 2.1(b). This is called breadth-first search as the order of exploring the nodes is a breadth-major fashion as shown in Figure 2.1(c).

2.3.4.2 Depth-first search

Let us consider the trees shown in Figure 2.2(a,b and c) generated by applying several applicable rules to its different nodes. The sequence of generation of the nodes in this case is different from that in Figure 2.1(c). Figure 2.2(a) is the state generated by applying rules to the start node A. The next step is to apply all the applicable rules to the node B and generate all the successor nodes as shown in Figure 2.2(b). In the next step, all the applicable rules are applied to the node E. This process continues until the goal state is reached or there are no more applicable rules in which case the search process backtracks to explore other nodes. The difference between the breadth-first and depth-first search lies in the sequence of exploration of the nodes as they are generated. The sequence of exploring nodes in a depth-first search is shown in Figure 2.2(c).

In some cases, depth-first search may prove more efficient. For example, when there are many solution paths leading to the solutions, but each one of them is quite long. In some cases, both approaches prove inefficient and may lead to *combinatorial explosion.*

2.3.4.3 Best-first Search

Best-first search is a combination of advantages of both depth-first and breadth-first search. At every step, the node with the highest potential is selected to be explored. This is achieved by applying heuristic functions to each one of them. In the following step, all the applicable rules are applied to the chosen node and its successors generated. One of the generated nodes could be the solution. If not, all those new nodes are added to the set of nodes already generated. Again, the most promising node is selected and the process continues until a solution is found.

2.3.5 Directions of Reasoning

Any search process could be carried out in the following two directions:

1. Forward - starting from the initial states; and

2. Backward - starting from the goal states.

Forward reasoning, or chaining as it is usually called, involves starting the reasoning process from the initial state and finally ending up at the goal state. The following rule is an example of forward-chaining:

> "if the span is less then 60 metres then a single span portal frame is an alternative."

The same rule could be reversed in the following manner:

> "Single span portal frame is an alternative if the span is less than 60 metres"

This latter rule is an example of a backward-chaining reasoning. As is clear from this example, backward-chaining involves starting from a goal state and proving the initial state by working its way backwards. The relative merits of the two approaches is beyond the scope of this discussion and can be found in (6).

2.3.6 Weak Methods

The search methods mentioned in Section 2.3.4 are basic search techniques and may be used by or within other problem-solving methods commonly referred to as *weak methods* for the reasons that will be mentioned in Section 2.6.1. The following are brief descriptions of some of the frequently used problem-solving methods used in Artificial Intelligence.

2.3.6.1 Generate and test

The generate and test method is probably one of the simplest control strategies used in AI. A typical generate and test strategy consists of the following steps (2,4,6):-

1. generate a likely solution;
2. test or confirm the solution generated in the previous step; and
3. if the solution is right, stop or start again from step 1.

The most important drawback of this approach is that if the solution space is very large, it might take a very long time to find the solution. However, this strategy is bound to find a solution if one exists. Generate and test uses a depth-first search.

2.3.6.2 Hierarchical Planning and Least Commitment Principle

In certain domains (such as design) the whole problem has to be solved by developing a plan of solution at different abstraction levels. The problem is divided into different subproblems at successive levels of abstraction. The detail increases as one progresses from any level to the one below it. In this manner, the problem is hierarchically decomposed into *loosely coupled subproblems*. As the solution process progresses, a number of solutions might exist to any subproblem. At the same time, a solution to one subproblem may depend upon decisions made in the others. One approach to this problem is to defer the binding decisions. This deferment of binding decisions is called the *least commitment principle* which effectively means the postponement of giving values to variables until more information is available.

2.3.6.3 Top Down Refinement

Some types of problems have to follow a particular plan for their solution. In design particularly, this normally involves moving from the top to the lowest level through different levels. Every level in the hierarchy is a refinement of the level immediately above it. Thus, the solution process involves passing from the most to the least abstract level.

2.3.6.4 Constraint Satisfaction

Many problems require determining values of different parameters of some entity to ensure that they satisfy different constraints. Design falls under this category. Constraint satisfaction does not require new search methods on its own and uses any of the basic search methods discussed in Section 2.3.4. Stefik (8) extended the classical constraint satisfaction method by integrating it into hierarchical planning. He called it constraint posting and divided the process into three distinct stages:-

1. constraint formulation - which is the operation of adding new constraints as the solution progresses;
2. constraint propagation - which is the operation of creating new constraints from the old ones as a result of interactions between subproblems; and

3. constraint satisfaction - which is the operation of finding values to different parameters such that the constraints are satisfied.

2.3.6.5 Backtracking

In essence, backtracking involves retracing back the solution path once it is found to be infeasible. In its simplest form, backtracking involves retracing back the steps followed so far in the solution process. This is called chronological backtracking. Chronological backtracking is one of the basic facilities provided in PROLOG (9,10). In another form of backtracking called the *dependency-directed backtracking* (DDB), developed by Stallman and Sussman (11), a database is built up which keeps a record of the following:

1. all deduced facts during the course of the solution process;
2. all the antecedent facts of the deduced facts;
3. the support justifications of the facts; and
4. the relevant rules.

These are known as dependency records. Support justifications are justifications for any assumption made during the solution process. The dependency records are used to determine the antecedents to be withdrawn whenever a wrong solution is identified. One obvious drawback of DDB is that it requires a lot of book-keeping. However, this extra work can be fruitfully used for generating explanations. DDB is used for *non-monotonic reasoning* described below.

Non-monotonic Reasoning

Non-monotonic reasoning involves a reasoning in which a new conclusion may invalidate some previous conclusions. This means that the *set of beliefs* may undergo changes as new conclusions are drawn by a system. Conversely, in a *monotonic* system conclusions keep adding on and there is no check on the consistency of the set of statements held true by the system at any time. Thus, *monotonic* reasoning quite obviously lacks a fundamental property of *set of beliefs*, i.e., consistency. *Non-monotonic reasoning* accomplishes this by keeping a record of all the *supports* of any conclusion and testing for their presence whenever a new conclusion is drawn. If a new conclusion invalidates the support of some previous conclusion, that conclusion is also invalidated, thus, maintaining a consistent *set of beliefs*. *Non-monotonic reasoning*, quite obviously, is better suited to dealing with situations that may arise in real domains. The following are some common situations for *non-monotonic reasoning* (6,12):

1. if a system has incomplete knowledge, it may have to make default assumptions which must be invalidated when more knowledge is added;
2. when the situations are constantly changing;
3. when temporary assumptions may have to be made in complex problem-solving.

2.4 Truth Maintenance System

Doyle (1) utilized the dependency-directed backtracking to develop the Truth Maintenance System (TMS). The TMS is a general system which may be used by other inferencing systems to check the consistency of statements generated by it. Whenever an inconsistency is detected by the TMS, it evokes its reasoning mechanism (which uses DDB) to resolve it by altering a minimal set of statements generated.

TMS consists of two basic data structures called *nodes* and *justifications*. *Nodes* represent a statement or a rule. The state of a node will be one of the following:-

IN - a node believed to be true at a particular moment; and

OUT - a node not believed to be true at a particular moment either because:

(i) no reasons exist to believe it, or

(ii) none of the possible reasons are currently valid.

A *justification* represents a way of establishing the validity of a node. Every node has a list of justifications attached to it. Quite obviously, IN node has at least one currently valid justification and an OUT node has none. There are two kinds of justifications in TMS as follows:

Support-list (SL) - (SL(in-nodes) (out-nodes))
Conditional Proof (CP) - (CP<consequent-node>(in-hypothesis) (out-hypothesis))

An SL justification is valid if all the nodes in the in-nodes are currently IN and all those in the out-nodes are OUT. A CP justification is valid if the consequent node is always IN whenever the nodes in the in-hypothesis are IN and those in the out-hypothesis are OUT.

Whenever an inconsistency is detected a CONTRADICTION node is added with the appropriate SL list. At this point, the DDB mechanism is evoked to resolve it. This is accomplished in the following manner. This is now considered with respect to the following example having the nodes 1, 2 and 3 in the beginning with the corresponding justifications:

(1) Object is a bird (SL(2)(3))
(2) Object has wings
(3) Object has an engine

These statements effectively say that "if an object has wings and there is no evidence that it has an engine then conclude that it is a bird". After some reasoning, a fourth node is added with appropriate justification:

(1) Object is a bird (SL(2)(3))
(2) Object has wings
(3) Object has an engine
(4) Object has undercarriage (SL (8,12) ( ))

At this point, it is concluded that nodes 1 and 4 cannot be true simultaneously. This is conveyed to the TMS by creating a fifth CONTRADICTION node:

(5) CONTRADICTION (SL(1,4)( ))

This justification simply means that the contradiction has occurred because of nodes 1 and 4 being true at the same time. The DDB mechanism is invoked at this stage. This involves tracing back through the nodes in the SL justifications for the

CONTRADICTION node and then through the nodes in their justifications and so on. In order to resolve the contradiction, TMS tries to find a set of inconsistent assumptions and then eliminate them. In this example, node 1 is an *assumption* - an assumption is a statement which has a non-empty OUT-list. This is recorded by creating a NOGOOD node with a CP justification as follows:

(6) NOGOOD N-1 (CP 5 (1,4)( ))

Now TMS chooses one of the inconsistent assumptions and makes it OUT by making one of the nodes in its OUT list IN. In this case there is only one assumption (i.e., node 1) and there is only one node in its OUT list (i.e., node 3). So, TMS makes Node 1 OUT by making node 3 IN. At this stage, the set of nodes becomes:

(1) Object is a bird (SL(2)(3))
(2) Object has wings
(3) Object has an engine (SL (6)( ))
(4) Object has undercarriage (SL (8,12) ( ))
(5) CONTRADICTION (SL(1,4)( ))
(6) NOGOOD N-1 (CP 5 (1,4)( ))

The moment node 3 becomes IN, node 1 becomes OUT as it depended on node 3 being OUT. Since node 1 becomes OUT, node 5 also becomes OUT as it depended on node 1 being IN. Thus, the contradiction is resolved. One noteworthy point is that node 4 (object has undercarriage) still remains IN as it was not involved in the contradiction and, thus, the reasoning required to reach that conclusion need not be repeated.

The difference between SL and CP justification also becomes clear from this example. To explain the difference, node 6 is given an SL justification instead of a CP justification as shown above. In this case, the justification for node 6 would be the following:

(6) NOGOOD N-1 (SL (5) ( ))

In this case, node 6 would depend solely on node 5 being IN. But that would mean that when node 5 becomes OUT, node 6 would become OUT too. This would lead to node 3 becoming OUT as well. The moment node 3 becomes OUT, node 1 would again become IN, thus, leaving the original contradiction unresolved. Many details of the operation of TMS have been omitted for this discussion but the description presented here gives an overall view of how TMS works.

2.5 Knowledge Representation

Knowledge representation is one of the central issues in the development of any AI program. The most important aspect of any AI system is the proper definition of the relationships between the different *symbols* used in the program. Various kinds of formalisms have been developed to represent any knowledge. These are a collection of relationships between *symbols*. In the following paragraphs, a brief discussion of the different commonly used formalisms is presented.

2.5.1 Production Rules

Production rules (12,13) are probably the oldest formalism of all. They have been in existence since the 1940s. A production rule is just a statement in the following form:

"if these conditions hold, then do something"

ie., a condition-action rule. The conditions in a production rule normally refer to the presence or absence of elements in the working memory.

2.5.2 Semantic Nets

The term *semantic network* (12,13) represents a network in which the nodes typically represent objects, concepts or situations and the arcs the relationships between them. For example, to represent the fact that 'John is a bright student', we create the following network with the corresponding two nodes and a link between them:

john -->isa-->bright_student

We could add the fact that 'a bright student is an intelligent person':

john -->isa-->bright_student -->isa-->intelligent_person

'To deduce new facts from these facts, a program just needs to have rules about intelligent person'.

2.5.3 Frames

Frames (12,13) are an extension of the idea of semantic networks that were proposed by Minsky (14) in 1975. Frames are a way of grouping information in records made up of *slots*. The value of a *slot* is called its *filler*. Hence, frames are categorised under *slot-filler* formalisms. For example:

```
[frame 10: portal_frame
 [isa: structure]
 [material:
  range:(steel concrete)
  default: steel]]
```

Frames are normally organised into a hierarchy to allow inheritance of properties. There are many special-purpose languages designed to handle frame-based representation, eg., ART, KEE (20).

2.6 Knowledge-Based Systems

2.6.1 Emergence of Knowledge-Based Systems

As mentioned in Section 2.3, the problem-solving methods discussed earlier are categorised as *weak methods*. Laird and Newell (15) define a *weak method* as:

> *"a problem-solving method that makes limited demands for the knowledge of the task environment and provides a schema within which domain knowledge can be used."*

Although these methods have been quite successfully used to solve problems (eg. game playing, etc.) requiring 'intelligence', their power was found to be quite limited in so far as solving practical complex problems is concerned. One of the chief reasons for calling them weak methods is that they may lead to *combinatorial explosion* as the complexity of problems increases. Another important reason was that most of the problems these methods solved were 'common sense' reasoning tasks, ie., they do not require any special kind of knowledge to solve. However, there is a whole array of tasks that require specialist knowledge to solve, eg., engineering design, medical diagnosis, etc. This realisation was the result of years of research in AI. Consequently, some researchers changed their focus of attention and turned to working on application problems in well-defined, narrow domains.

Perhaps the first public presentation of some insights into knowledge-based experts systems (KBESs) was given by Feigenbaum (16) at the 1977 International Joint Conference on Artificial Intelligence (IJCAI), paraphrased here,

> *"the power of an expert system derives from the knowledge it possesses and not from the particular formalisms and inference schemes it employs."*

However, in spite of the empirical observations that an expert's knowledge per se seemed both necessary and nearly sufficient for developing an expert system, few researchers were still working towards comprehensive knowledge representation theories and associated general-purpose systems. But sooner than later it was realised that these efforts had limited powers for pretty much the same reasons as weak methods. The central role that knowledge played in solving knowledge-based problems with the performance levels of an expert was also soon realised. In Feigenbaum's words (18),"the expert's knowledge provides the key to expert performance, while knowledge representation and inference schemes provide the mechanism for its use."

Figure 2.3 displays chronological developments of some of the major works of expert systems. The success of some of these systems (eg., DENDRAL, PROSPECTOR, MYCIN etc.) quite obviously aroused a great deal of interest in expert systems, and methods employed in these systems were soon categorised under *strong methods*. The main component of these systems responsible for such a categorisation was the domain-dependant knowledge that they possessed.

Based on the characteristics of the successful systems like DENDRAL and PROSPECTOR, there were many attempts to define what is an expert system. However, just as the definition of AI, there are numerous definitions of expert systems proposed by different researchers, some too ambitious, some vague and some theoretical. The following is a definition of expert systems proposed by Gasching (17) that seems most apt,

> *"Expert systems are interactive computer programs that solve problems that, if solved by humans, would require intelligence and expertise."*

2.6.2 Types of KBESs

Knowledge-based expert systems have been developed in many areas covering a wide range of applications. All these applications can be broadly classified into the following categories:

1. diagnosis;
2. design;
3. data interpretation;
4. planning; and
5. education.

The following table (16) shows the different applications of expert systems:

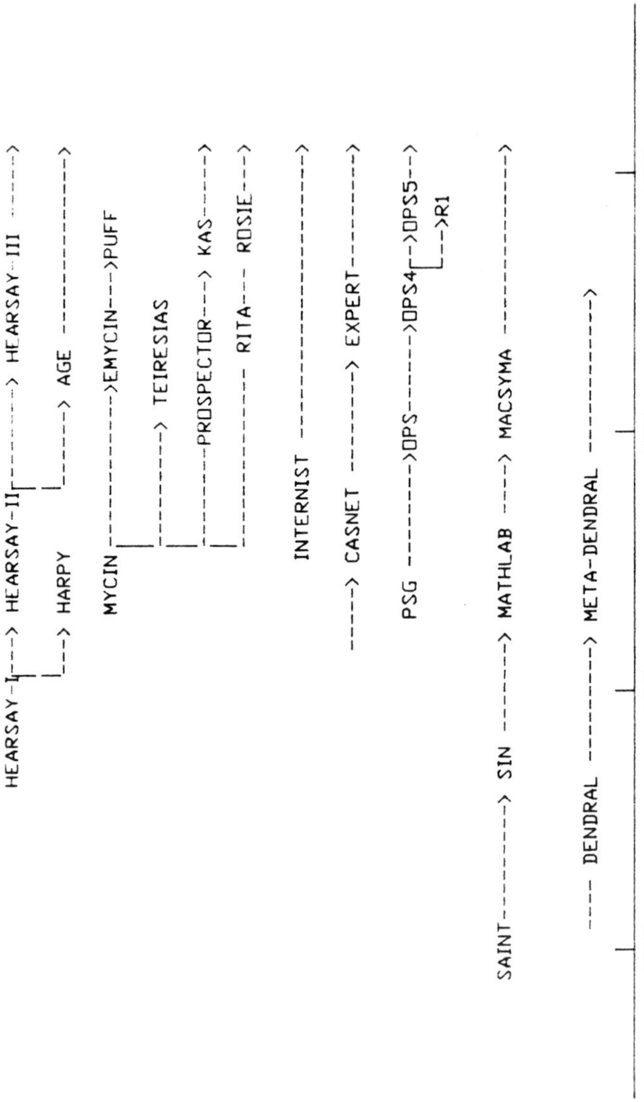

Figure 2.3 Reproduced from (16)

Category	Problem Addressed
Interpretation	Inferring situation descriptions from sensor data
Prediction	Inferring likely consequences of given situations
Diagnosis	Inferring system malfunctions from observables
Design	Configuring objects under constraints
Planning	Designing actions
Monitoring	Comparing observations to plan vulnerabilities
Debugging	Prescribing remedies for malfunctions
Repair	Executing a plan to administer a prescribed remedy
Instruction	Diagnosing, debugging and repairing student behaviour
Control	Interpreting, predicting, repairing and monitoring system behaviours

2.6.3 Components of a KBES

There are three main components of a knowledge-based expert system. They are:

1. the knowledge base;
2. the inference engine; and
3. the user interface.

In addition to these, there could be other components in some systems like the knowledge acquisition and explanation facility. Figure 2.4 (18) is a schematic diagram of a typical knowledge-based expert system. The knowledge base is the main store house of information of the system. In some systems, the knowledge base may consist of declarative as well as procedural knowledge. The procedural knowledge is sometimes known as the control knowledge or meta-knowledge (ie., knowledge about knowledge). The inference mechanism decides the strategy about running the system, this involves the sequencing and firing of the rules.

2.6.4 Difference between a Conventional Program and a KBES

It is difficult to point out the exact differences between existing conventional programs and the KBESs. It is a matter of considerable debate to prove whether a particular KBES is really something different from a conventional program. However, the following differences can be stated (18) from a theoretical point of view and a *genuine* knowledge-based expert system may be said to have the following attributes:

1. KBESs are knowledge-intensive programs;
2. KBESs use rules of thumb to improve the efficiency of search;
3. KBESs have transparent knowledge bases and it is, generally, easy to expand the knowledge base;
4. in a KBES, there is, usually, a separation between the knowledge and the methods of manipulating that knowledge; and
5. KBESs can explain their reasoning behind reaching a particular solution.

2.7 AI tools

2.7.1 Introduction

One of the primary considerations in the development of knowledge-based systems is the selection of the appropriate development tool. The many options before a developer may be classified into the following categories:

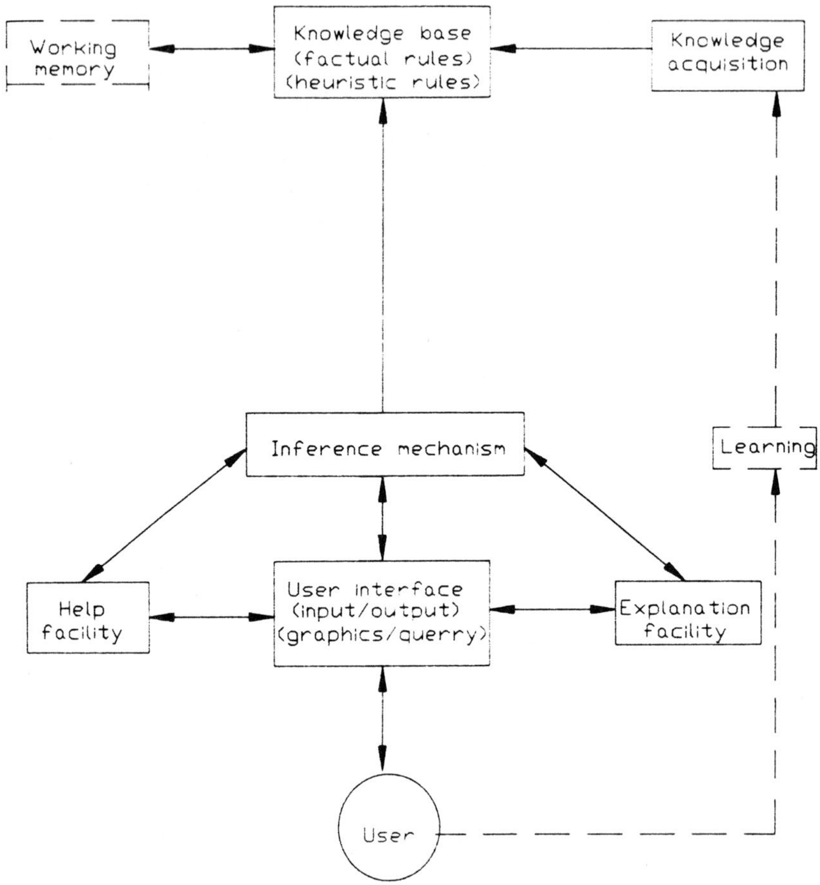

Figure 2.4 Components of an expert system
(reproduced from (18))

1. programming languages;
2. expert system shells; and
3. expert system development environments.

It is fast becoming a formidable task to choose a particular tool most suited for a particular requirement. Recent years have seen rapid multiplication of KBES development tools which leave many researchers thoroughly perplexed and confused. Sections 2.7.1.1 and 2.7.1.2 briefly describe two of the popular tools. Section 2.7.2 describes in some detail the particular tool used in this project, viz. the Edinburgh Prolog Blackboard Shell (24).

2.7.1.1 ART

ART (19) is an acronym standing for Automated Reasoning Tool. It was developed by Inference Corporation and is one of the more powerful tools implemented in LISP. There are four main components of ART:

1. a knowledge language;
2. a compiler for transforming the knowledge into LISP;
3. an inference system called the 'knowledge applier'; and
4. a development environment.

ART uses production rules, object-attribute-value triplets, 'facts' in a logical -relation notation and a procedural representation for algorithmic processes for knowledge representation. Its inference mechanism, or the 'knowledge applier', employs both forward and backward chaining strategies.

2.7.1.2 KEE

KEE (20) is another acronym standing for Knowledge Engineering Environment. It was developed by the IntelliCorp and is implemented in LISP (21). It is a multi-formalism environment as it provides the facility of representing knowledge as production rules or frames. Its inference mechanism consists of both forward and backward chaining. Any procedural knowledge written in LISP can also be attached to any slot of any frame. It has a very comprehensive graphics facility and most of the user interface is accomplished through graphics. It can display the whole line of reasoning graphically as tree structures, However, KEE is one of the most expensive AI tools currently available.

2.7.2 Edinburgh Prolog Blackboard Shell

2.7.2.1 Introduction

The development tool used in this project is called the Edinburgh Prolog Blackboard Shell (EPBS). EPBS (22) is written in PROLOG and supports the development of blackboard architecture systems. The development of EPBS was based on earlier blackboard systems particularly the HEARSAY-II (23) speech recognition system. The EPBS was originally intended to be developed as a research tool on user modelling (24). It is still an experimental tool and its development continues. The following sections describe the EPBS in some detail to enable the appreciation of its use in later chapters. Before embarking on a formal description of the EPBS, a general introduction to blackboard systems is given.

2.7.2.2 Simple blackboard systems

Blackboard architecture (25) is intended to support development of systems in domains characterized by interaction between diverse sources of knowledge. The blackboard serves as a global data structure which facilitates this interaction. A common analogy may be made to problem-solving in domains where a number of experts in different areas of specialities co-operate over the solution which any one of them could never achieve alone. In order to facilitate this process, they agree to use a blackboard to post (or write) any partial result they can contribute separately. Each expert takes turns to write on the blackboard and, in case more experts wish to write simultaneously, the conflict is resolved by some pre-defined strategy.

The main components of a typical blackboard system are:

1. entries;
2. knowledge sources; and
3. blackboard.

Entries

Entries are the immediate results produced by the system. In a typical system, each entry has a certainty factor as well as a specific identification.

Knowledge source

Knowledge sources (KSs) contain the knowledge embodied in the system. KSs contribute to the creation of entries posted on the blackboard. In a production rule system, KSs may be thought of as a collection of 'if...then...' rules in which the 'if-part' (or condition) of each rule, typically, refers to the presence or absence of some entry on the blackboard and the 'then-part' (or action) suggests some action to be taken resulting in some changes to the blackboard entries.

Blackboard

The blackboard is a global data structure whose main function is the organisation of entries and thereby handling the communication between the different knowledge sources. Typically, a blackboard is partitioned into a number of parts each representing different aspects or stages of the solution process.

Apart from the three components of a blackboard system described above, there is another equally important component called the *agenda.* An agenda is a list of KSs or rules to be executed in the next cycle. Based on the success or failure of a particular rule, new rules may get added on to it or some may be deleted from it. The basis of giving priorities to the rules on the agenda (ie., scheduling) may vary from system to system. Some simple system may apply some fixed criteria but in more sophisticated systems the scheduling, too, may be embedded in the reasoning process of the system itself.

2.7.2.3 Syntax of rules in the EPBS

The EPBS is a forward-chaining production rule based system. A typical rule in the EPBS syntax has the following form:

if	Condition
then	Goal
to	Effect
est	Est.

The following sections briefly describe each part of the rule.

Condition

'Condition' of a rule is a combination of tests on the blackboard, which essentially consists of testing the presence or absence of certain kinds of entries on the blackboard. A test is defined as follows:

Cf_test:= X<N;X>N;(N<X,X<M).
Atomic_test:=[Index_pattern,Fact_pattern,Cf_test]
Test:=Atomic_test;not Atomic_test; notnow Atomic_test;holds Precondition.
Condition:= Test;Test and Condition;Test or Condition.

where,

Index_pattern and Fact_pattern are PROLOG terms;
X is a variable;
N and M are terms representing certainty factors;

Index_pattern and Fact_pattern are PROLOG terms chosen by the user to identify a particular entry on the blackboard.

Goal of a rule

Goal is a PROLOG procedure to be executed by each rule. This gives the user the flexibility of only partially specifying an entry when writing the rule and further specification comes from the success of the condition or by calling the Goal. In case no other specification is required other than that prescribed by the condition, the PROLOG goal *true* may be called which will succeed immediately.

Effect of a rule

The effect of a rule could be one of the following:

add[Index,Fact,Cf], which adds an entry Fact on the blackboard under the index Index with certainty factor Cf,
or amend[Index,Fact,Cf], which ammends an entry Fact on the blackboard under the index Index with certainty factor Cf,
or action Action, which takes an action Action,
or delete, which deletes an entry from the blackboard.

where Index, Fact, Cf and Action are PROLOG terms.

Est refers to the usefulness of a rule. By default simple integers may be assigned to *Est* in which case the *usefulness* of the rules increase as the value of their Ests. This default rule can be over-ruled by any arbitrary PROLOG term to Est and these can be compared by a predicate defined by the user. *Est* provides a way of resolving conflicts regarding firing of rules. In cases where more than one rule are eligible to be fired due to the success of their conditions, the rule with the lowest *Est* is fired first, then the one with the next higher *Est* and so on.

Entry in the EPBS

An entry on the blackboard of the EPBS has the following form:

bb(Tag,Status,Index,Fact,Cf)

where Index and Fact are PROLOG terms and Cf is the certainty factor. Tag is an identification number assigned to every entry by the system. Status is assigned to every entry to indicate its state at a particular moment and could be one of the following:

1. in;
2. amended; or
3. inout.

Any entry may have one of these statuses depending on the cause of their presence on the blackboard. An entry which is simply added in the blackboard is assigned the status 'in'. All atomic tests are carried out on only those entries having status 'in'. An entry amended as a result of success of a rule is given the status 'amended'. These entries are not considered by an atomic test (discussed in Section 2.7.2.4). A 'not' test-marker (also discussed in Section 2.7.2.4) supporting an entry of the same form as the 'not' test itself is given the status 'inout'. Index provides a way of partitioning the blackboard. Different entries corresponding to different parts of the solution process may be grouped together under different indices. Any entry on the blackboard is unique and is identifiable using one or more of the arguments of the above-mentioned predicate bb/5 ('bb' is the name of the predicate and '5' its number of arguments).

Front-end facilities of EPBS

EPBS provides a number of front-end commands to facilitate the development process. The following is a list of some of them:

(1) show this - displays the current cycle,
(2) show next - displays the next cycle,
(3) show brief - displays a brief summary of the current cycle,
(4) show changes - displays the entries that went 'in' or 'out' of the blackboard,
(5) show agenda - displays the current agenda of rules to fire,
(6) show rule(Integer) - displays the rule number Integer,
(7) show rules - displays all the rules,
(8) show entries - displays all the entries currently on the blackboard.

2.7.2.4 Consistency Maintenance in the EPBS

Consistency maintenance of the blackboard is very important whenever affecting any change to it, ie., while adding or deleting an entry. Consistency maintenance becomes particularly important in the presence of 'not' tests. It is easy to conceive the addition of an entry invalidating a 'not' test. To accomplish this, the system keeps a record of all the 'not' tests tried that were part of a successful condition. This is done by adding a 'not' test-marker of the following form:

bb(Tag,in,Index,not_test(Fact,Cf_test,Rule_no),sure)

In case an entry is added to the blackboard that invalidates a 'not' test, the 'not' test is removed as well as everything it supports. However there is one exception. There can be a rule with a condition having a 'not' test about a certain kind of entry and the conclusion adding an entry of the same form. In this case the 'not' entry is invalidated by the very entry it supports! In this case, the status of the 'not' test is amended to 'inout' merely to indicate that they have been so affected.

Another typical need for consistency maintenance arises when an entry is supposed to be amended. In this case, the old entry which is to be amended no longer remains valid after the addition of the new amended entry and should be removed. However, it is needed to support the new entry otherwise if it is removed the new entry will be removed as well. Thus, to overcome this problem, it is still kept in the blackboard but with a different status. Its status is changed to 'amended' from 'in', which will be the status of the new amended entry. Such an entry (with 'amended' status) cannot be used by any atomic test. It is present as a support of the new entry and to explain how the new entry came about.

The final case considered for consistency maintenance is to test whether the addition of an entry does not invalidate some previously successful 'not' test that indirectly supports it. This actually is a case of a set of rules giving rise to a loop. In this case, a simple loop check is performed and if such a loop is found, the proposed entry is not added to the blackboard.

2.7.2.5 Some General Comments on the EPBS

One of the most important characteristics of the EPBS is the flexibility it provides for switching over to PROLOG programming at the user's will. This is accomplished by the following two features of the shell:

(i) provision of the 'holds' test;
(ii) provision of the PROLOG Goal in the 'then' part of a rule; and

(iii)provision of a PROLOG term as the Action.

The provision of the 'holds' test enables one to accomplish very complex pattern matching in the left hand side or the condition of a rule. Using 'holds', one can have very powerful general conditions to be tested by encoding the condition as a PROLOG procedure. For example, the following rule has a lengthy PROLOG procedure as its condition:

```
if    holds structure_type(X)
then   true
to    add[structure_type,X,true]
est   5.
```

This rule uses the 'holds' test in its condition and represents the fact that if structure type is X then add an entry on the blackboard that the structure is of type X, where 'structure_type(X)' is a PROLOG procedure which could be quite complex and lengthy.

The provision of a PROLOG Goal in each rule provides the flexibility to jump out of the 'if-then' syntax of a rule and program the desired operation in PROLOG as and when appropriate. If the rule syntax proves to be too rigid in some cases, the code can be written completely in PROLOG. Thus, EPBS provides all the features of PROLOG alongwith the additional features of the shell.

Another useful feature of EPBS is the user's control over the scheduling system. By defining one's own strategy and giving appropriate values to Est, the user has complete control over the firing sequence of the rules. In sophisticated systems, Est may even have a variable value which may be evaluated by the system's reasoning process itself.

Because of the features described above, EPBS is a fairly powerful tool particularly for research purposes. However, on the debit side is the fact that it does not have any graphic facilities. It also does not have a multi-formalism environment which might be useful in many cases.

2.8 Summary and Conclusions

Brief descriptions of some of the important problem-solving methods in Artificial Intelligence were presented. Some of the popular artificial intelligence tool were briefly reviewed. A detailed description of the particular tool used in this project was also given which included the description of some of its powers as well as its limitations. It is, however, concluded that none of the tools could be found to be tailor-made for any particular purpose. But, at the same time, any tool could serve the purposes of most applications, to a certain extent. This leads to the conclusion that the time spent in exploring and evaluating the different tools commercially available for one's own particular requirements is not well-spent! The final lesson learnt from this project, as regards the selection of appropriate tools, is that most tools are suitable for most problems, the only difference being that certain features might be more easily incorporated in a system using one more than the others. In the beginning, the EPBS did not seem to offer the features being looked for but gradually as the project developed, it was felt (as illustrated later on) that it had a lot of potential.

References

1. **Doyle, J.**, "*A Truth Maintenance System*", Artificial Intelligence, Vol. 12, No. 3, pp. 231-272, 1979.

2. **Nilsson, N.J.**, "*Problem-solving Methods in Artificial Intelligence*", McGraw Hill Book Company, 1971.

3. **Boden, Margaret**, "*Artificial Intelligence and Natural Man*", The M.I.T. Press, London, England, 1987.

4. **Winston, P.H.**, "*Artificial Intelligence*", M.I.T. Press, Cambridge, Mass., 1978.

5. **Charniak, E. and McDermott, D.**, "*An Introduction to Artificial Intelligence*, Addison-Wesley Publishing Company, 1985.

6. **Rich, Elaine**, "*Artificial Intelligence*", McGraw Hill Book Company, 1986.

7. **Minsky, M.**, "*Semantic Information Processing*", M.I.T. Press, Cambridge, Mass., 1968.

8. **Stefik, M.**, "*Planning with Constraints (MOLGEN: Part I)*", Artificial Intelligence, Vol. 16, pp. 111-140, 1981.

9. **Clocksin, M.F. and Mellish, C.S.**, "*Programming in Prolog*", Springer-Verlag, Berlin, 1987.

10. **Bratko, J.**, "*Prolog Programming for Artificial Intelligence*", Addison-Wesley Publishing Company, 1986.

11. **Stallman, R.M. and Sussman, G.J.**, "*Forward Reasoning and Dependency-directed Backtracking in a System for Computer-aided Circuit Analysis*", Artificial Intelligence, Vol. 9, No. 2, pp. 135-196, 1977.

12. **Frost, R.A.**, "*An Introduction to Knowledge Base Systems*", Collins, London, 1986.

13. **Ross, Peter**, "*Expert Systems Course for M.Sc./Ph.D. - 1985/86*", Department of Artificial Intelligence Teaching Paper No. 1, University of Edinburgh, 1985.

14. **Minsky, M.**, "*A Framework for Representing Knowledge*", The Psychology of Computer Vision, Ed. P. Winston, McGraw Hill, New York, 1975.

15. **Laird, J. and Newell, A.**, "*A Universal Weak Method*", Report No. CMU-CS-83-141, Department of Computer Science, Carnegie-Mellon University, Pittsburgh, U.S.A., 1983.

16. **Hayes-Roth, F, Waterman, D. and Lenat, D. (Eds.)**, "*Building Expert Systems*", Addison-Wesley Publishing Company, page 6, 1983.

17. **Gasching, J. et. al.**, "*Development of a Knowledge-based System for Water Resources Problems*", *SRI Project 1619, SRI International, Palo Alto, California, August 1981.*

18. **Adeli, Hojjat**, "*Expert Systems in Construction and Structural Engineering*", Chapman and Hall, 1988.

19. **Clayton, B.D.**, "*ART-Programming Tutorial, Vols. 1-3*", Inference Corporation, Los Angeles, California, U.S.A., 1985.

20. **Lurent, J., et. al.**, "*Comparative Evaluation of Three Expert Systems Development Tools: KEE, Knowledge Craft and ART*", The Knowledge Engineering Review, Vol. 1, No. 4, pp. 18-29, 1986.

21. **Winston , P.H. and Horn, B.**, "*LISP*", Addison-Wesley Publishing Company, Reading, Mass., 1981.

22. **Johnson, K. and Chan, N.**, "*Edinburgh Blackboard Shell User's Manual*", Artificial Intelligence Applications Institute, University of Edinburgh, 1987.

23. **Erman, L.D. et. al.**, "*The HERASAY-II Speech Understanding System: Integrating Knowledge to resolve uncertainty*", Computing Surveys, Vol. 12, No. 21980, pp. 213-253, 1980.

24. **Milington, M, and Jones, J.**, "*The Edinburgh Prolog Blackboard Shell*", Department of Artificial Intelligence, University of Edinburgh, 1986.

25. **Hayes-Roth, Barbara**, "*A Blackboard Architecture for Control*", Artificial Intelligence, Vol. 26, pp. 251-321, 1985.

Chapter 3

Literature Review

3.1 Introduction

In recent years, interest in the research into the application of Artificial Intelligence tools and techniques to Structural Engineering problems has rapidly developed. It is, however, still in its infancy and we have yet to see the use of these methods in practice properly demonstrated. This chapter reviews some of the research work and takes a close look at the emphasis of current research efforts.

Before reviewing the systems, a scheme for the classification will be outlined. The current research may be classified into the following two broad categories:

(i) the knowledge-based approach; and
(ii) the analytical or numerical approach.

As the name suggests, the knowledge-based approach is used in the development of knowledge-based systems. A knowledge-based system may or may not be an expert system depending on whether or not it incorporates expert knowledge and possesses the different characteristics of an expert system (as outlined in Section 2.6). If it does, it may be termed a *knowledge-based expert system* (KBES) otherwise simply a knowledge-based system. However, both these names are used interchangeably and it becomes difficult to ascertain the exact connotations in different contexts. In fact, for some reason, the general tendency is to call any artificial intelligence program (particularly in the application domains) a *knowledge-based expert system* or simply an *expert system.*

Apart from knowledge-based approaches, there have been a few attempts at developing *intelligent* systems using *analytical* or *numerical* methods. These approaches have mainly centred on feedback mechanisms in conventional programs. There is, generally, absolutely no knowledge manipulation involved which is why they are not considered to be knowledge-based systems. However, they are also applications of Artificial Intelligence in that they are attempts at simulating the learning process. The typical approach of this type of work (22-24) has been to store all the successful results and to subsequently utilise them by making comparisons between the problem at hand and those solved in the past. Another approach in this category was that used in the development of DESIGNER (1). DESIGNER is an *intelligent* system for ship building design. It uses the dependencies between different design 'entities' to infer the effect of some change in one or more of them on the others. It may also be used to explain its reasoning to some extent. However, it can not be called a knowledge-based approach as there is no manipulation of a knowledge-base. It does have a knowledge-base but the manipulation is numerical, rather, than symbolic.

In the following paragraphs, a brief discussion of different existing and proposed systems are given.

3.2 Knowledge-Based Systems in Structural Engineering

3.2.1 Knowledge-based Approach

HI-RISE (4) is a knowledge-based system for the preliminary design of high-rise buildings. It is implemented in a frame-based language, PSRL. The knowledge-base of HI-RISE contains declarative as well as procedural knowledge. The declarative knowledge represents a physical hierarchy of known structural types. The procedural knowledge is organised into 'knowledge modules'. The top level modules concern the design of two functional systems: a lateral load resisting system and a gravity load resisting system. Each of these systems is further decomposed into different knowledge modules. Its inference mechanism is that used by PSRL (5). The knowledge-base in HI-RISE comprises of production rules grouped in rule sets, a collection of schema templates and several Lisp functions. Production rules are used to represent heuristic elimination rules concerning the synthesis of alternative configurations. Schemas are used to represent *feasibility constraints* and *evaluation constraints*. The production rule constraints are formulated and statically stored in the knowledge-base whereas the schema constraints are formulated dynamically during the solution of a particular problem. Following are a couple of elimination rules taken from HI-RISE used for the synthesis of lateral load resisting systems:

If the number of stories > 50
AND 3D system is core
THEN alternative is eliminated.

IF 3D system is tube
AND 2D system is solid-shear wall
THEN alternative is eliminated.

One of the major handicaps of HI-RISE is that it has a fixed agenda. This means it cannot handle situations that demand a different sequence of operations. Another limitation is that its context tree built during the consultation, can only represent rectangular buildings; it is unable to represent buildings of other geometries. However, HI-RISE was a major contribution in the sense that it has successfully demonstrated the application of a knowledge-based approach to structural design and was successful in addressing important issues such as the representation of design information and the choice of a design strategy.

SACON (6) which stands for Structural Analysis CONsultant is an early example of a knowledge-based system in structural engineering. It was implemented using the EMYCIN (6) system. The knowledge base of SACON consists of :

1) rules for inferring analysis strategies indicating the most appropriate analysis and the associated recommendations,
2) rules for inferring the critical stress, deflections and other behaviour of structures, and
3) mathematical models for estimating non-dimensional stress and deflection constraints on each substructure based on boundary and loading conditions.

In SACON, the deep knowledge is represented as four-tuples made up of an associative triple (object-attribute-value) and its certainty factor. The heuristic rules are represented as production rules. The following is an example rule:

```
Premise: ($And(Same Cntxt Material(list of metals))
        (Between* Cntxt Error 5 30)
        (Greaterp* Cntxt No-Stress 0.9)
        (Between* Cntxt Cycles 1000 100000)
Action : (Conclude Cntxt SS-Stress Fatigue Tally 1.0)
```

The English translation of this would be:

If

1. The material composing the substructure is one of the metals and
2. The analysis error is between 5% and 30% and
3. The non-dimensional stress of the substructure is greater than 0.9 and
4. The number of cycles of the applied loading is between 1000 and 100000.

Then

it is definite (1.0) that fatigue is one of the stress phenomena in the substructure.

The control strategy in SACON is backward-chaining and performs a depth-first search. SACON was mainly developed to act as a front-end to a finite element analysis program, MARC. SACON divides the structure into substructures and each substructure is dealt with separately. The overall response of the structure is determined from the peak responses of the substructures. The final recommendation of the system is the most appropriate analysis strategy for the problem at hand.

SESCON (7) is another prototype system developed on the lines of SACON to be used as a front-end to the Seasam-69 structural analysis package.

Garrett (8,9) has developed a knowledge-based standards processor, SPEX. It is proposed that this processor would act as an interface between CAD programs and the design standards. This processor may be utilised for either designing or checking structural components for conformance with a design standard and other external constraints. The standards are represented as a network of decision tables (10) in SPEX. Figure 3.1 shows the schematic model of SPEX. One of the major features of this system is that the standard-dependant and the standard-independant knowledge are separated. This has the following advantages:

(1) Changes in the design standard may be dealt with separately without affecting the CAD program;
(2) The interpretation and the formal representation of the design standard has to be undertaken once only and not every time a CAD program is written.
(3) The CAD program may be used for any standard by simply changing the standard-dependant knowledge which the standard processor uses to satisfy the CAD program's requests.

The knowledge base of SPEX consists of the following knowledge modules :-

(a) Design focus generation module - This KM generates a hypothesis which is later translated into standard requirements by the requirement retrieval

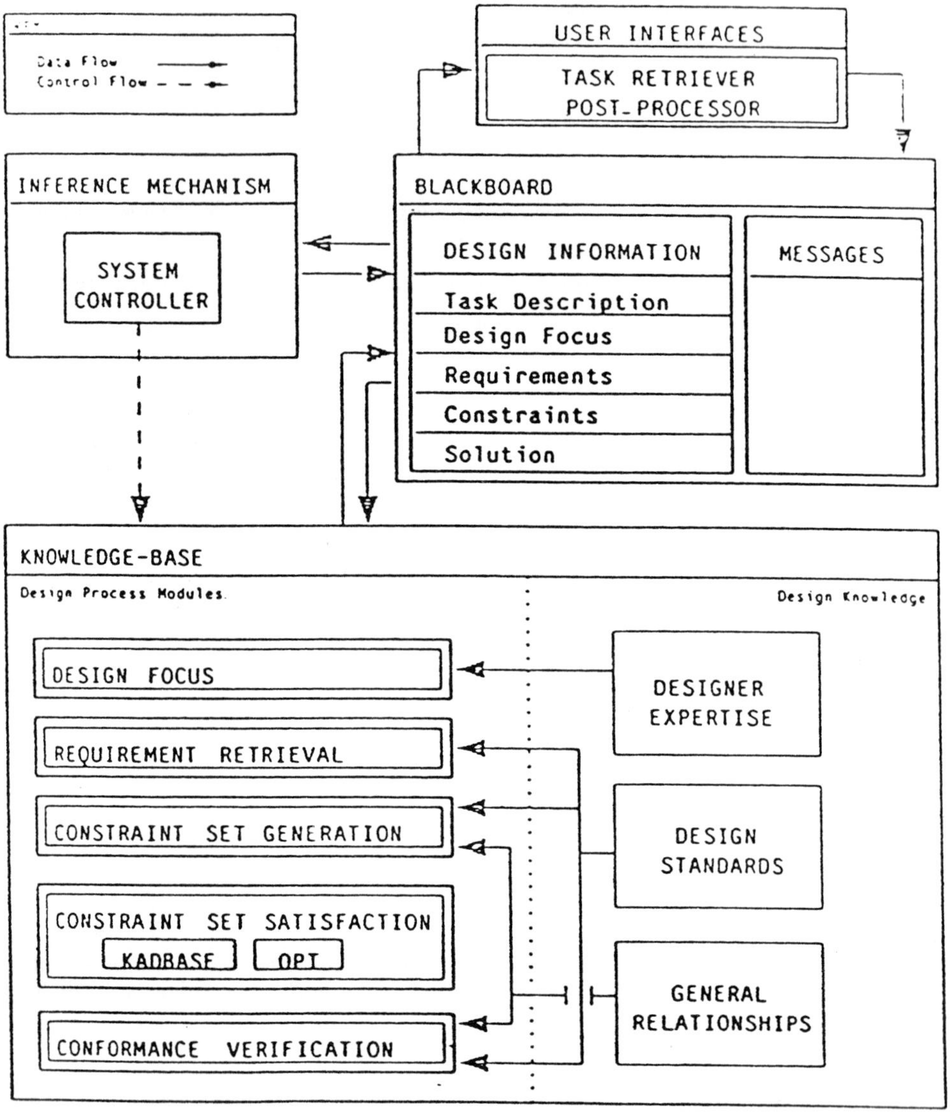

Figure 3.1 Schematic model of SPEX (taken from (8))

module.

(b) Requirement retrieval module - This KM retrieves the standard requirements.

(c) Constraint set generation module - This KM generates a set of constraints that when satisfied, guarantees that the requirements on the requirement satisfy list are satisfied.

(d) Constraint set generation module - This KM retrieves the standard requirements.

(e) Constraint set satisfaction module - This KM is responsible for finding an optimum solution to the set of constraints built by the constraint generation module and the external constraints, if any.

(f) Conformance verification module - This KM is responsible for checking all requirements found in the requirement check list built by the requirement retrieval module and placing any violated requirements in the requirement violated list.

One of the handicaps of SPEX is that it may only be used to design components of a structure and cannot be used for the complete structural design. Another handicap of SPEX is its over-reliance on its optimisation routine. Whenever its optimisation routine cannot find a solution, it immediately assumes that either the constraint set was incorrect or the hypothesis was wrong instead of considering the fact that the optimisation routine itself may have failed to converge to a solution. However, SPEX is probably to date the most comprehensive example of knowledge-based standards processor developed.

SPECON (11) is a small prototype knowledge-based system intended to assist the designer in checking structural steel elements for conformance with the AISC (12) steel design specifications. It was developed at Carnegie-Melon University as a part of a class project. Its organisation is similar to that of MYCIN and consists of:

(1) the knowledge base;
(2) the context; and
(3) the inference machine.

The knowledge-base of SPECON is divided into two levels, one of which identifies the applicable constraints, whereas the other comprises the specific design constraints. The context is a Short Term Memory that keeps track of the various facts generated in a particular consultation. The control strategy in the inference machine is similar to MYCIN's backward-chaining strategy. The provision of a code in SPECON is represented as production rules. It also has an explanation module which can answer questions relating to:

(1) how a certain hypothesis was deduced; and
(2) why a certain question was asked.

SPECON is a very small prototype system and has only one clause of the AISC specifications in its knowledge base. It is, however, a good demonstration of the application of the production rule approach to standards processing.

Sriram (13) proposed a conceptual model for the integrated structural design called DESTINY. It integrates all the stages of the structural design process into a unified framework. Its scope is limited to the design of buildings. DESTINY utilises a blackboard system of architecture. Figure 3.2 is a conceptual view of the DESTINY

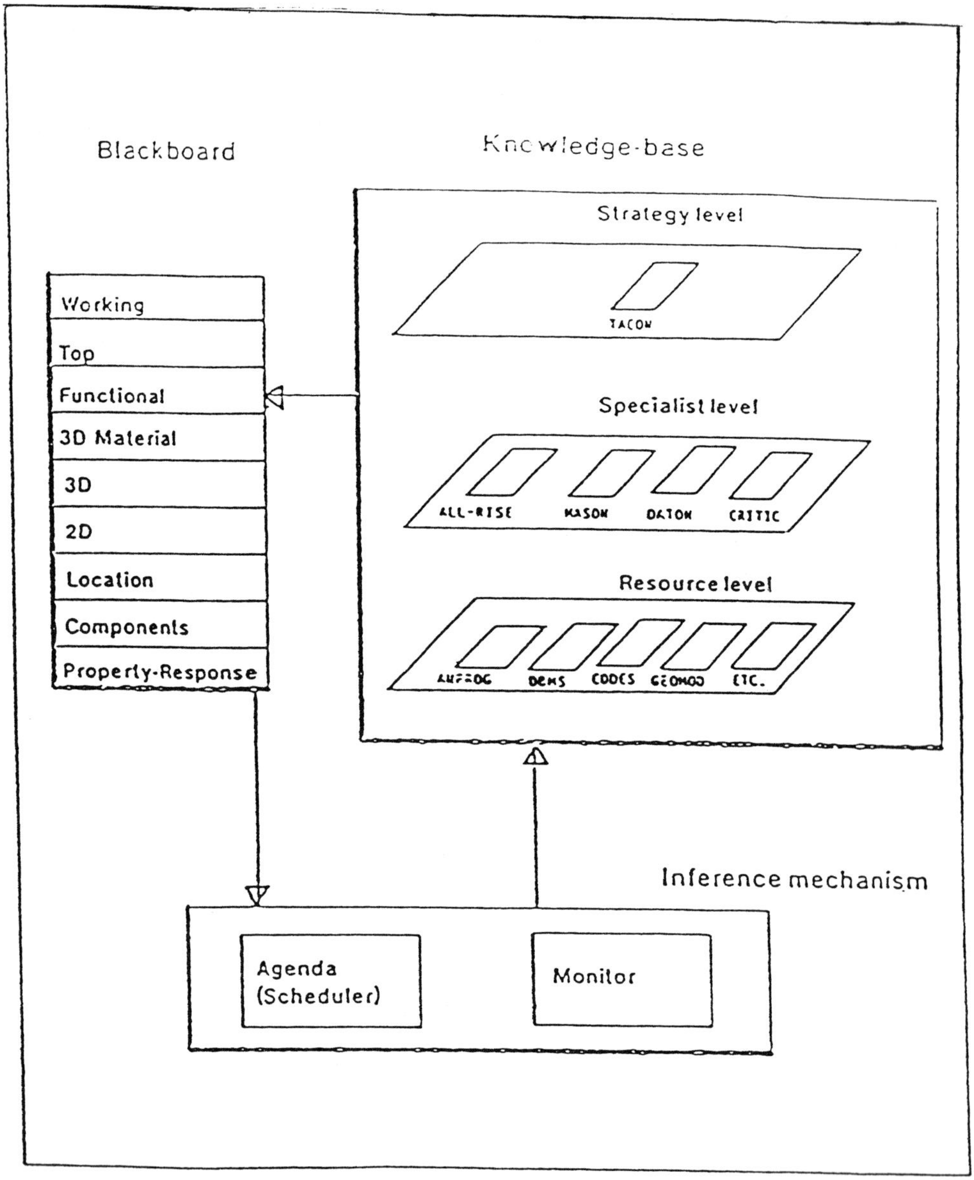

Figure 3.2 Schematic model of DESTINY (taken from (13))

system. The knowledge base of DESTINY consists of a number of different knowledge modules. These are organised into a hierarchy of three levels:

1. Strategy level knowledge modules : These modules analyse the current solution state to determine the next course of action. In the reported version, only one knowledge module - TACON - exists at a level. This level schedules the execution of the specialist level knowledge modules.
2. Specialist level knowledge modules : These modules contribute to the development of the solution on the blackboard. Most of the knowledge modules at this level are themselves small KBSs, having a knowledge base consisting of engineering heuristics. The different knowledge modules at this level are:

   (i) ALL-RISE : This knowledge module synthesises different alternatives from the input related to space planning. This is the only knowledge module implemented so far. Its knowledge base consists of schemas representing the static knowledge hierarchy and schemas representing the constraints that determine feasible alternatives. The building design knowledge is stored in schemas, production rules and LISP functions. It has approximately 100 LISP functions, a few OPS5 rules, 270 schema templates.
   (ii) MASON : This knowledge module models and analyses the feasible structural configurations.
   (iii) DATON : The knowledge module proportions and details linear and surface structural elements such as beams, columns and the like.
   (iv) CRITIC : This knowledge module criticises and evaluates the current best design.

   It is at this level that detailing and planning is undertaken by the DATON module.

3. Resource level knowledge modules : These contain the analytical knowledge and reference information required for analysis and design. The knowledge modules at this level consist mainly of algorithmic programs and database management systems (DBMS).

The inference mechanism of DESTINY has two main components :

(1) Agenda; and
(2) Monitor.

The strategy level knowledge module schedules the execution of the specialist level tasks and sets the Agenda. The monitor then executes the element with the highest priority.

Of all the knowledge modules of DESTINY, ALL-RISE is the only one to be implemented so far. As already stated, it synthesises a number of feasible alternative structural systems from the input provided after the space planning. ALL-RISE can itself be considered to be a separate knowledge-based expert system. It is an extension of HI-RISE (4) to include buildings (residential or commercial) of any number of stories. The knowledge base of ALL-RISE consists of the following:

(1) a static hierarchy of schemas that represents the abstraction hierarchy of the building; and
(2) a constraints knowledge base that represents feasible combinations of sub-systems and feasible alternatives.

The inference mechanism of ALL-RISE consists of the following steps:

(1) Based on the input which is the spatial planning of the building, a solution tree consisting of a number of feasible alternatives is generated using top-down refinement and constraint handling. The alternatives are generated using a depth-first processing. This is accomplished by:
    (a) the synthesis constraints pruning the solution tree;
    (b) deferring commitments by posting interaction constraints between sub-systems, if sub-systems taking part in the constraint have not yet been generated.
(2) The solution process is terminated at a pre-determined level in the *static knowledge hierarchy* (SKH). At this level, all the feasible alternatives are extracted and evaluated.
(3) The best evaluated design, ie., the current best design is posted on DESTINY's blackboard.

Adeli and Al-Rijleh (14) have developed a knowledge-based expert system for the design of roof trusses called RTEXPERT. It can advise on the appropriate type of roof truss, selection of the layout of the truss and the loading. By the layout of the truss it is meant the pitch of the truss and the number of panels in the truss. RTEXPERT has been developed using the INSIGHT 2+ (15) expert system shell. RTEXPERT can also compute the nodal forces due to various loads. All these computations are undertaken using routines written in Turbo Pascal. It is not clear from the literature whether RTEXPERT is an expert system or not. The problem solving methods used are not clearly expressed either. It is, thus, difficult to comment upon the 'intelligence' of the system.

Rasdorf and Wang (16) have developed a knowledge based standards processor called SPIKE. SPIKE is an acronym standing for Standards Processing In a Knowledge based Expert system environment. It can carry out mainly two tasks, viz.

1. design conformance checking; and
2. determining allowable value ranges for undetermined design datums.

The main difference between SPIKE and other standards processors lies in the representation of the standard provisions. SPIKE uses a 'factual representation of standards'. A standard is represented as provisional and organisational facts rather than rules. The basis of the facts is the decision table model of standards. The implementation language used is OPS5. The generic nature of SPIKE is accomplished by a generic processor which manipulates the factbase (knowledge base) of SPIKE. The processor can manipulate any standard represented using the schema adopted by SPIKE.

Gero (17) and his group at the Computer Applications Research Unit of the University of Sydney have been actively involved in the development of knowledge-based systems for design synthesis. The basic issue addressed in this group's works is the applicability of inference systems to generate design. It is proposed that the design synthesis is the reverse of design analysis and could be accomplished by reversing the process of design analysis (18). By design analysis it is meant the evaluation of a particular design, ie., the evaluation of the performance of a given object. The 'design descriptions' may be interpreted by proper inferencing techniques to derive certain properties of design. The question tackled is, "Can the interpretation of design 'specifications' produce design 'descriptions'?" It is considered that the design synthesis is the interpretation of a design specification. The reason being that in the synthesis, one starts with a description of a required performance or a specification and ends up with a description of an object. Thus, it is shown that the same knowledge-base and architecture used for design analysis may be used for design synthesis. An example of applying this idea to design is clear from Figure 3.3. This figure is a part of a run of a program used by Rosenman (19) based on the BUILD shell (19). In this example the user wants to know why the fire rating was **not** 'none'. To reply, the system has to search through the knowledge base to find out all the conditions when the fire rating requirement would be 'none' which means that given the required specifications, the system can find out the required object description. In this particular example, it has found the 'object description' in the following three parts:

(1) the fire resisting construction must be type 5;
(2) the building classification must be type 1; and
(3) the building must be a house.

This is a simple example but certainly has the potential to be applied to more complex cases.

Based on the above principles, a system for the design of retaining walls, RETWALL (20), was developed. RETWALL selects the most appropriate retaining wall prototype for the problem at hand and then sizes it for the imposed design loads. It reduces the range of options by making a judicious inquiry about soil and topographical conditions as well as designer preferences. This example is classified at the derivation end of the problem-solving spectrum where the solution is derived from a set of pre-defined solutions. This approach will produce results only if the number of options to be considered is relatively small. However, if there are an innumerable number of options to be considered, it is not a straight mapping between the specifications and design descriptions as discussed earlier. For problems falling under this category, Coyne and Gero propose more models based on the concepts of planning and multiple abstractions (21).

Coyne (21) developed a knowledge-based system for the spatial layout design of a room based on the use of production rules which are generative rather than being directly inferential as in RETWALL. Looking from the point of view of problem-solving strategies adopted, RETWALL uses a 'derivation' approach whereas the room layout design system uses a 'formation' approach. When it is talked about the 'generation of designs' from the 'interpretations of specifications', it essentially is the 'formation' approach very similar to the one adopted for HI-RISE (4).

In a nutshell, the work of the Computer Applications Research Unit of the University of Sydney concerns the following two ideas:

(i) interpretations of design specifications to produce design descriptions; and
(ii) from interpretation to generation of design.

Gero and his colleagues (25,26) have also undertaken some work on knowledge-based approaches towards optimization in structural design.

```
enter command
*****************************************************************
?- explain why_not 'fire resistance rating required' is_ none
*****************************************************************

fire resisting construction_ type 5
   needed to prove
fire resistance rating required is_ none


building classification is_class I
   needed to prove
fire resisting construction is_type 5


building is_ house
   needed to prove
building classification is_ class I
```

Figure 3.3 Part of a run of a program using the BUILD shell (reproduced from (19))

3.2.2 Analytical approach

Rooney and Smith (22,23) were probably one of the first to apply AI techniques to structural design. Their approach was, however, not a knowledge-based one. Instead, they experimented with the introduction of a feedback mechanism in a conventional design program by developing an expanded model of the design process based on flow of information. Figure 3.4 is a representation of the algorithm, thus, developed. The concept behind the introduction of the feedback mechanism was to use past results to produce solutions to problems at hand. It was, in some ways, an attempt at developing a 'learning' program.

The feedback mechanism is composed of the following three steps :

(1) acquisition of experience;
(2) application of experience; and
(3) database management.

The key feature of this research was to develop a system capable of self-modifying its operation and change its decisions based upon previous successes and failures. It was quite a comprehensive effort at developing an 'intelligent' system for structural design. However, there were certain important issues not addressed, eg., the treatment of 'soft' constraints. Since the 'soft' constraints play such a vital role in practical designs, it is difficult to evaluate the actual value of this work from the point of view of its utility to the industry.

Jozwiak (24) discusses a very similar approach to that of Rooney and Smith (23) for improving the effectiveness for structural optimization programs. The main objective of his work was to reduce the computing time since that is one of the most important factors in optimisation programs as the design cycles or iterations for any real-life structure is very large. Here again, the key element of the research was to take advantage of previous design results for the selection of initial values of the decision variables and rearrangement of the constraints. Figure 3.5 is the flow chart of the program, thus, developed (24) utilising the feedback mechanism.

Successful results were stored (ie. when optimum was found) in a data file. All the unsuccessful results and near-misses were neglected. These successful results were used in later problems by comparing the structures designed in the past and the one being designed.

The program developed using these concepts showed a marked improvement over the conventional programs in terms of computing times. On average, the 'intelligent' program having the feedback mechanism required 2.1 times less number of iterations than the conventional program to find an optimum solution.

MacCallum (1,27) discusses a different approach to design based on a network model of design. His contention is that a satisfactory approach to using computers for engineering design should satisfy the following requirements:

(a) interactive control of the design process by the designer;
(b) flexibility to add new design parameters which require to be satisfied;
(c) flexibility to define new relationships to be used and the conditions under which they would be evaluated;
(d) assistance in calculations; and
(e) feedback on information on direct and indirect relationships between parameters.

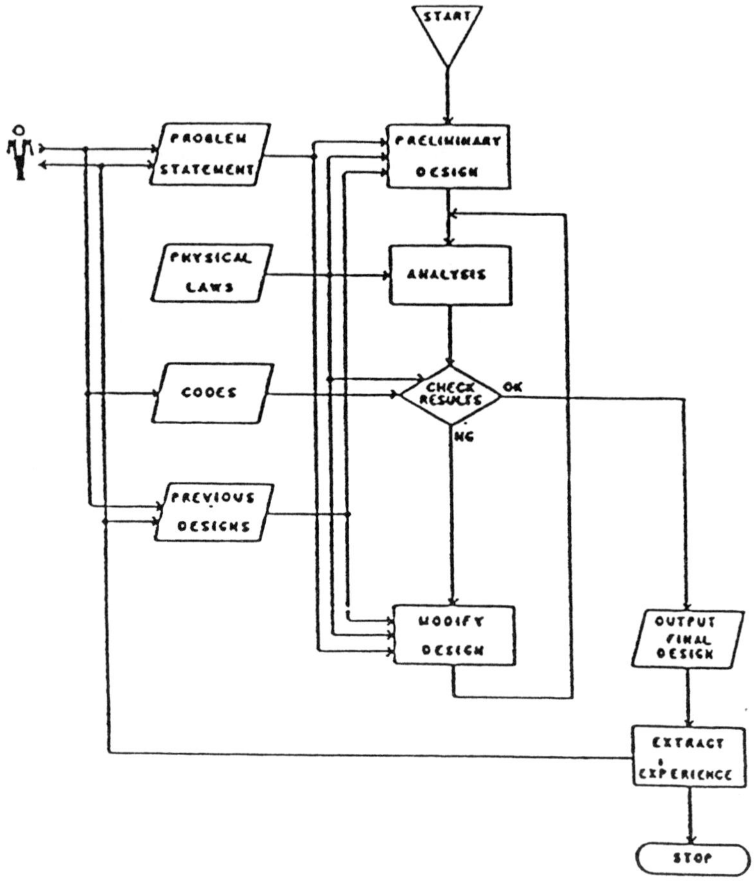

Figure 3.4 The algorithm developed for utilising feedbacks from past designs (taken from (23))

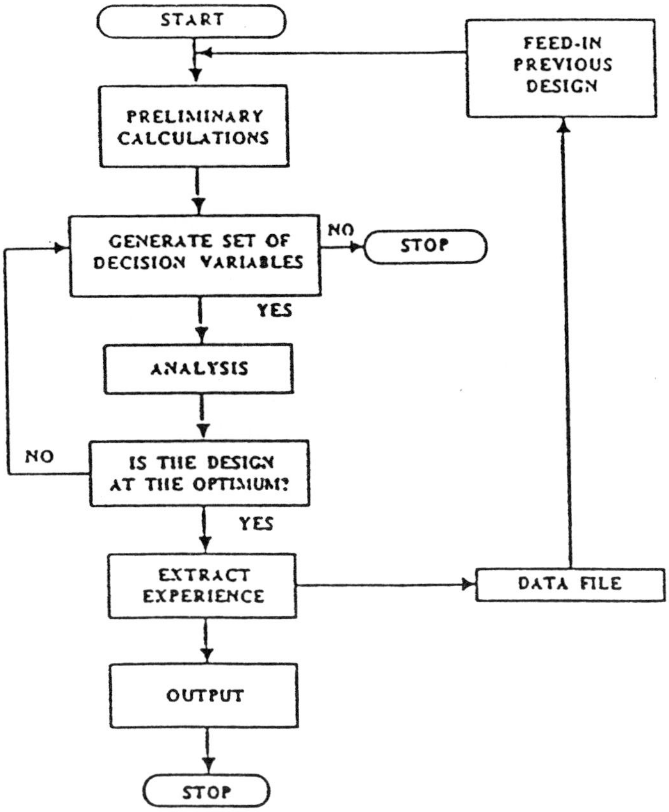

Figure 3.5 The algorithm developed for utilising feedbacks from past results of an optimisation program (taken from (24))

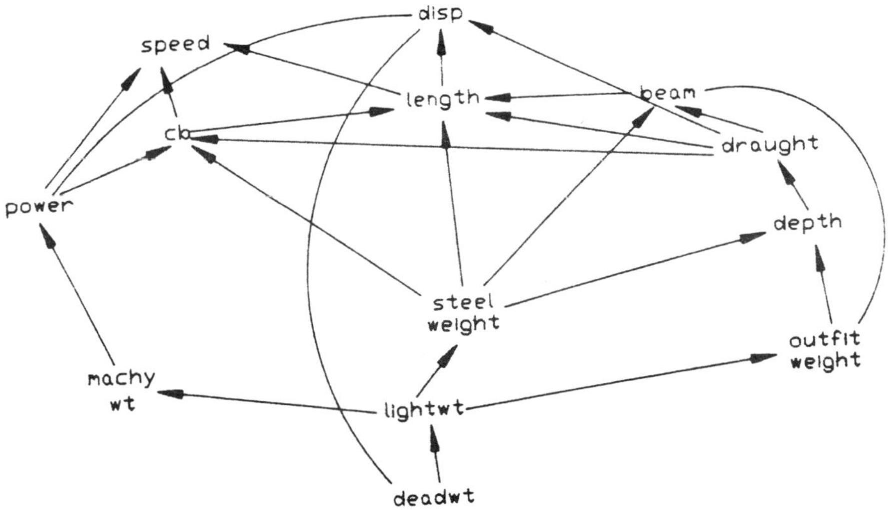

Figure 3.6 A network for preliminary ship design (reproduced from (1))

MacCallum considers that one of the most challenging problems in design is the awareness and understanding of the influence of one characteristic or parameter of the system being designed on the others. These relationships could be either social, numerical, geometrical or spatial. Figure 3.6 is a network for preliminary ship design. Based on these ideas, a system was developed called DESIGNER. The most important feature of the system is that it allows a network to be created by the user. The user has complete control over the number of 'entities' to be used in the network. Different networks may be readily created or existing networks modified. Another important feature of DESIGNER is to provide information on the relationships between different parameters or characteristics. DESIGNER can also explain how it reached a particular conclusion.

The approach used in DESIGNER is a powerful one in the sense that it gives complete flexibility to the user to use or 'abuse' the system as one wishes to. This is a feature that the conventional systems lack. However, the DESIGNER system only uses numerical relationships and the manipulation is numerical. Incorporation of other major relationships would be a significant advance.

3.3 Final Comments

It is quite clear from the foregoing discussions of some of the systems that the general approaches of the current research is in the direction of either choosing alternative structural systems or detailed sizing and proportioning and checking of the members of a structure. However, in practice, these are not the issues that cause most difficulties for designers On the contrary, the problems that need experience and expertise to solve fall into a different category altogether. Once a structural system has been chosen and its members sized and proportioned and checked, there are numerous occasions when the designer has to change his design completely or utilise another structural system altogether. Some reasons for such a situation arising could be that:

(1) some 'local' constraints are not considered;
(2) some provisions of the design codes are mis-interpreted; or
(3) even some provisions of the design codes are ignored !

Other problems may arise due to some unforeseen interaction between different components of a structure. It is these problems that require *intelligence* and expert knowledge to solve and not the sizing of a member. Instead of building design synthesis systems, if efforts are made to build systems that can advise on steps to be taken in problems arising in the above-given situations by reviewing the whole or partial design, the results will probably be more fruitful. The main issues to be tackled are the treatment of the *soft* constraints, interpretations of design code provisions and handling the different types of knowledge encountered in structural design, eg., conflicting and/or fuzzy data.

References

1. **MacCallum, K.J.**, "*Understanding relationships in Marine Systems Design*", Proceedings of the First International Marine Systems Design Conference - Theory and Practice of Design, pp. 1-9, 1982.

2. **Buchanan, B.G.**, *"DENDRAL and META-DENDRAL: Their Applications"*, Artificial Intelligence, Vol. 11, pp. 5-24.

3. **Shortliffe, E.R. et. al.**, *"Rule Based Expert Systems"*, Addison Wesley, Reading, Massachusetts, 1984.

4. **Maher, M.L. and Fenves, S.J.**, *"HI-RISE: An Expert System for the Preliminary Design of High Rise Buildings"*, Report No. R-85-146, Department of Civil Engineering, Carnegie-Mellon University, Pittsburgh, U.S.A., 1985.

5. **Rychener, M. and Fox, M.**, *"PSRL: An SRL-based Production System"* , Robotics Institute, Carnegie-Mellon University, Pitssburgh, U.S.A., 1984.

6. **Benett, J.S. and Englemore, R.S.**, *"Experience using EMYCIN* Rule-Based Expert Systems, Eds. B.G. Buchanan and Shortliffe, E.H., Addison Wesley, Reading, Massachusetts, pp. 314-319, 1984.

7. **Fjelheim, R. and Syversen, P.**, *"An Expert System for Sesam-69 Structural Analysis Program Selection"*, Technical Report No. Cp-83-6010, Division for Data Technology, Computas, Norway, 1983.

8. **Garrett, J.H., Jr. and Fenves, S.J.**, *"A Knowledge-based Standards Processor for Structural Component Design"*, Report No. R-86-157, Department of Civil Engineering, Carnegie-Mellon University, Pittsburgh, U.S.A., 1986.

9. **Fenves, S.J. and Garrett, J.H., Jr.**, *"Knowledge-Based Standards Processing*, Artificial Intelligence in Engineering, Vol. 1, No. 1, pp. 3-14, 1986.

10. **Harris, J.R. and Wright, R.N.**, *"Organisation of Building Standards: Systematic Techniques for Scope and Arrangement"*, Building Science Series NBS BSS 136, National Bureau of Standards, Washington, D.C., March, 1980.

11. **Sriram, D., Maher, M.L. et. al.**, *"Expert System for Civil Engineering - A Survey"*, Report No. R-82-137, Department of Civil Engineering, Carnegie Mellon University, Pittsburgh, U.S.A., pp. 37-39, July, 1982.

12.*Specification for the Design, Fabrication and Erection of Structural Steel for Buildings*, American Institute of Steel Construction (AISC), Chicago, U.S.A., 1978.

13. **Sriram, D.**, *"Knowledge-based Approaches to Structural Design"*, Ph.D. Dissertation, Department of Civil Engineering, Carnegie-Mellon University, Pittsburgh, U.S.A., 1986.

14. **Adeli, H. and Al-Rijleh, M.M.**, *"A Knowledge-based Expert System for the Design of Roof Trusses"*, Microcomputers in Civil Engineering, H. Adeli (Ed.), Vol. 2, No.3, 1987.

15. **Level Five Research in Pascal**, *"INSIGHT 2+ Reference Manual"*, Level Five Research, Indialantic, Florida, U.S.A., 1986.

16. **Rasdorf, W.J. and Wang, T.J.**, *"Generic Design Standards Processing in an Expert System Environment"*, Journal of Computing in Civil Engineering, Proceeding of the American Society of Civil Engineers, pp. 68-87, Vol. 2, No. 1, 1988.

17. *Department of Architectural Science, Design Computing Unit, University of Sydney, Knowledge Engineering Abstracts*, March, 1988.

18. **Gero, J. and Coyne, R.**, "*Development in Expert Systems for Design Synthesis*", Expert Systems in Civil Engineering, Ed. C.N. Kostem, American Society of Civil Engineers Publications, New York, pp. 193-203, 1986.

19. **Rosenman, M.A. et. al.**, "*An Expert System for Design Codes and Design Rules*", Applications of Artificial Intelligence in Engineering Problems, Ed. D. Sriram and R. Adey, Springer-Verlag Publishers, Berlin, pp. 745-758, 1986.

20. **Hutchinson, P.J. et. al.**, "*RETWALL: An Expert System for the Selection and Preliminary Design of Earth Retaining Structures*", Knowledge-based Systems, Butterworths, Vol. 1, No. 1, pp. 11-23, December, 1987.

21. **Coyne, R.D.**, "*A Logical Model of Design Synthesis*", Ph.D. Dissertation, Department of Architectural Science, University of Sydney, 1986.

22. **Smith, Steven, E.**, "*Artificial Intelligence Applied to the Preliminary Design of Simply Supported Beams*", Master's Dissertation, Rensselaer Polytechnic Institute, Troy, New York, U.S.A., August, 1983.

23. **Rooney, M.F. and Smith, S.E.**, "*Artificial Intelligence in Engineering Design*", Computers and Structures, Pergamon Press, Vol. 16, No. 1-4, pp. 279 - 288, 1983.

24. **Jozwiak, S.F.**, "*Improving Structural Optimization Programs using Artificial Intelligence Concepts*", Engineering Optimization, Gordon and Breach Publishers, Vol. 12, No. 2, pp.155-162, 1987.

25. **Balachandran, M. and Gero, J.S.**, "*A Knowledge-based Approach to Mathematical Modelling and Optimization*", Engineering Optimization, Gordon and Breach Publishers, Vol. 12, No. 2, pp.91-116, 1987.

26. **Balachandran, M. and Gero, J.S.**, "*Use of Knowledge in selection and Control of Optimization Algorithms*", Engineering Optimization, Gordon and Breach Publishers, Vol. 12, No. 2, pp.163-173, 1987.

27. **MacCallum, K.J.**, "*Creative Ship Design by Computer*", Computer Applications in the Automation of Shipyard Operation and Ship Design IV, Eds. D.F. Roger et. al., North-Holland Publishing Company, pp. 55-62, 1982.

Chapter 4

A Model for Integrated Structural Design

4.1 Introduction

This chapter presents a model for structural design which forms the framework for the development of a knowledge-based system for structural design. Different prototypes developed as part of this work are the components of the model being presented here. This model is based on an earlier model called DESTINY, developed by Sriram (1) and proposes some enhancements to it. The basic concepts behind the model are the same as DESTINY. However, a few additions were reckoned to be important for an integrated structural design system and were lacking in the DESTINY model. The model being presented here will be called INDEX, which stands for INDustrial Building Design EXpert. The reason for calling it INDustrial Building Design EXpert is that industrial buildings are taken as examples to illustrate the model. However, it is suggested that the model is general enough for structural design. The model presented in this chapter forms an overall framework for the development of a knowledge-based system for structural design. Three components of the model were developed in this work and are described in detail in Chapters 5, 6 and 7.

Section 4.2 presents a description of the structural design process. Although DESTINY has been described earlier in Section 3.3.1, a detailed description will first be presented in Section 4.3. as the INDEX model is an extension of the DESTINY model. Section 4.4 discusses some of the limitations of the DESTINY model. Section 4.5 will describe the INDEX model. Section 4.6 is a comparison of the two models highlighting the differences between the two. Finally, Section 4.7 contains a summary and conclusions of this chapter.

4.2 Structural Design

4.2.1 The Process

A civil engineering structure may be defined as an entity which will withstand the imposed design loads and transmit them to the foundations. In doing so, the structure must fulfill certain engineering and other architectural constraints. The structural design process includes the proportioning and sizing of such a structure to ensure the appropriate levels of safety and serviceability specified in the various design documents such as codes of practice and building regulations. The whole design process may be divided into three distinct stages :

1. Preliminary design: In this stage, the functional requirements and constraints are synthesised into a preliminary design concept. This involves the selection of a potential structural configuration satisfying layout and spatial constraints. This stage frequently includes an approximate analysis to evaluate the response of the alternative candidate structures selected for further consideration.
2. Detailed design: This involves the detailed design of candidate structure chosen in (1) and consists of the following three sub-stages :

   a. structural analysis ;

   b. proportioning and sizing the structural members ; and

c. checking all the applicable design constraints.

This stage typically consists of several iterations between *analysis* and *proportioning and sizing* to ensure that all applicable constraints are satisfied with economy of design. Most of these constraints are specified in the applicable design codes. There may be a few external constraints as well, such as restrictions on the height of a structure. A large and significant deviation in the properties of the components assumed at the analysis and proportioning stages might necessitate another analysis-proportion and sizing-check cycle. This is typical of most design problems. The iteration continues until a satisfactory design is arrived at. In some cases, there may be a return to the preliminary design stage resulting in a revision of the chosen structural concept.

3. Design documentation: Detailing of the different components and preparation of the design documents.

4.2.2 Discussion

Figure 4.1 (2) shows the different stages of the structural design process as well as indicating the influencing factors (experience, heuristics etc.) at every stage. The important thing to note is the feedback that may become necessary at any stage. At any stage, the designer may be forced to go back to almost any earlier stage and reconsider his decisions made earlier. This aspect of design forms the most difficult part to be incorporated in computer programs.

The description above of the structural design process is only a description of the different stages in structural design. However, the more important aspect of structural design is the inference mechanism involved. As discussed in the previous paragraph, the designer may be forced to go back and reconsider his earlier decisions and even change it in certain circumstances (Figure 4.1). In terms of Artificial Intelligence (AI), it would be said that there can be recurring changes in the *current set of beliefs* throughout the design process. This may arise due to a sudden change of specifications or the emergence of new constraints possibly in conflict with the existing ones. What is being talked about here is a *non-monotonic situation.* A brief discussion on *non-monotonic reasoning* is given in Section 2.3.6.5. For detailed discussion on the different aspects of *non-monotonic* reasoning, refer to (3,4). A more detailed treatment of the *non-monotonic* nature of the structural design process and ways of tackling it is presented in Chapter 7.

4.3 The DESTINY Model

4.3.1 Architecture

The DESTINY model is based on a blackboard architecture (5). The sole purpose behind selecting this architecture is to facilitate the communication between the different experts involved in the structural design process, e.g., the architects, space planners, service engineers etc. A brief description of how it is accomplished in the blackboard architecture environment may be found in Section 2.7.2.2. Figure 4.2 is a schematic representation of the DESTINY model.

4.3.2 Blackboard

DESTINY's blackboard is divided into two parts. The first part is called the *Working level* which contains entries relating to the execution of the various knowledge modules (KMs). The second part is split into eight levels, viz., *Top, Functional, Material, 3D, 2D, Location, Components and Property-Response.* These levels contain entries relating to the different stages of the design process. These levels

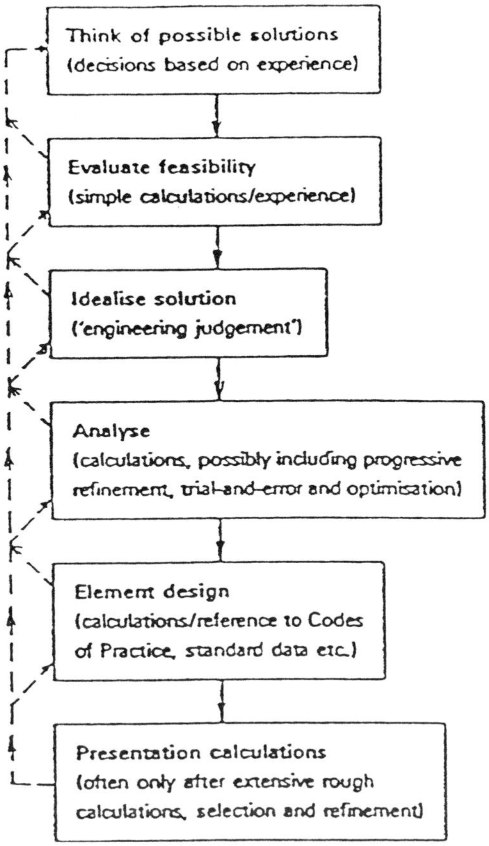

Figure 4.1 The different stages in structural design (taken from (2))

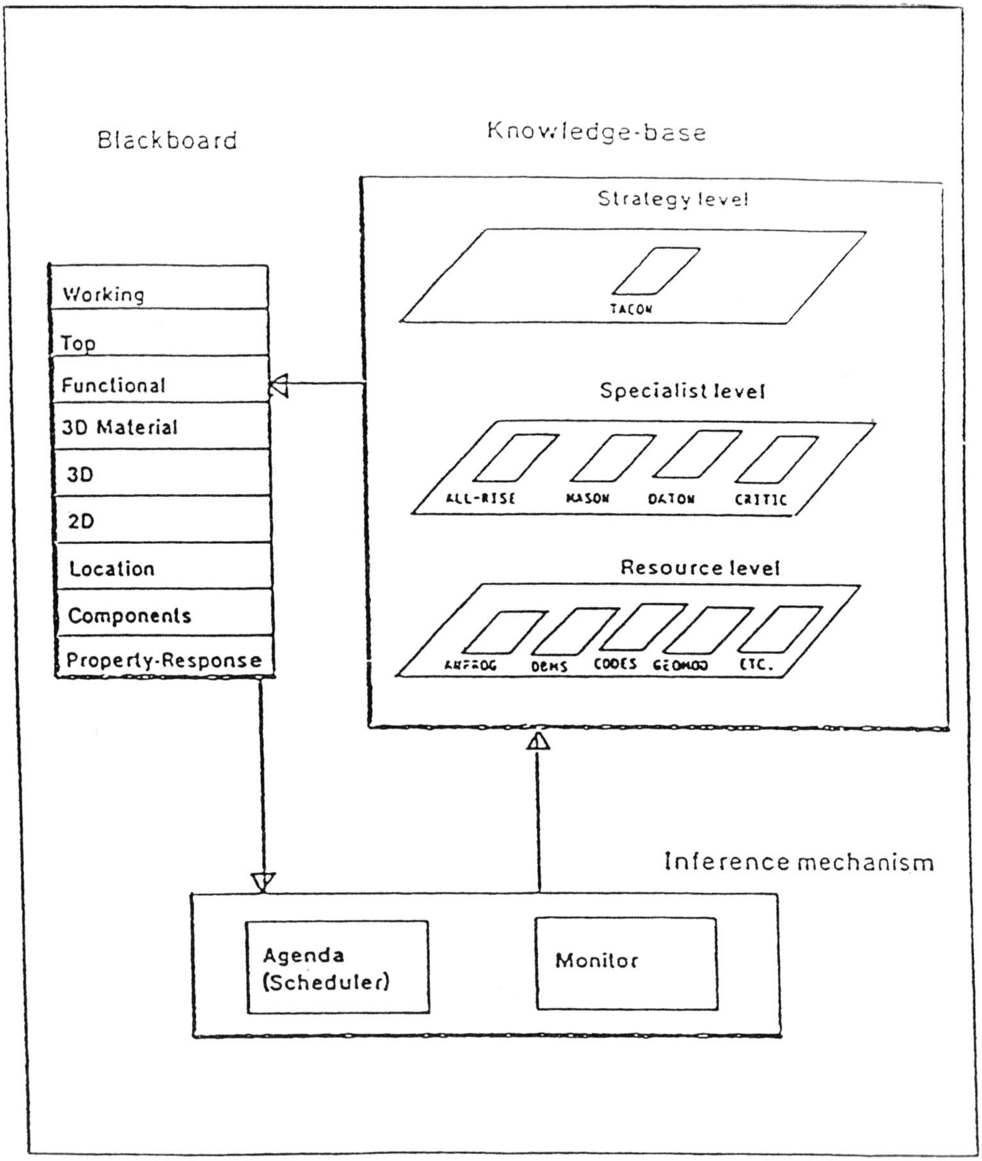

Figure 4.2 Schematic model of DESTINY
(taken from (1))

may be seen as a hierarchical decomposition of the building design process, i.e., they define the abstraction hierarchy of the design entities. Figure 4.3 shows the abstraction hierarchy on the blackboard of DESTINY.

4.3.3 Knowledge-Base

The knowledge-base of DESTINY is organised into a hierarchy of three levels. Each level comprises of a number of knowledge modules each performing a particular task in the design process. The three levels of DESTINY's knowledge base are:

1. the strategy level;
2. the specialist level; and
3. the resource level.

Brief description of these levels may be found in Section 3.3.1. Brief descriptions of the knowledge modules at these levels are also given in Section 3.3.1. A relatively detailed description of the modules relevant to the discussions of this chapter will be presented here. The module that is directly relevant to the extension of the DESTINY model proposed by this work is CRITIC. The purpose of this module is to check whether a design generated by the other modules (i.e., ALL-RISE, MASON and DATON) is satisfactory to perform the intended functions. This task of CRITIC is divided into two sub-tasks, viz., *Criticize and Evaluate*. The purpose of the *Criticize* sub-task is to assign one of the following four values to a design:

1. unsatisfactory;

2. modifiable;

3. fixable;

4. satisfactory.

There are four sets of production rules that determine one of these values for a particular design. The different case when a design is assigned one of these values is discussed below:

1. *Unsatisfactory* - this value is assigned to a design if the intended behaviour of the structure is not achieved.
2. *Modifiable* - this value is assigned to the design if there are significant differences between the assumed and the computed properties of the structure. A *modifiable* design is recommended to undergo a re-analysis.
3. *Fixable* - this value is assigned to the design if there are minor differences between the assumed and the computed properties of the structure. A *fixable* design is one which does not have to undergo a re-analysis and needs just minor adjustment to some of its parameters.
4. *Satisfactory* - this value is assigned to the design if it satisfies all the specifications previously laid down.

Once a design is found to be *satisfactory* then a detailed evaluation is carried out by CRITIC as the *Evaluate* sub-task.

For the descriptions of other knowledge modules, reference should be made to Section 3.3.1.

4.3.4 Inference Mechanism

As with any other blackboard system, the inference mechanism of DESTINY also consists of an *agenda* and a *monitor*. The *agenda* contains a list of the sequence of *specialist level* knowledge modules to be executed from the elements of the following set:

{ALL-RISE MASON DATON CRITIC}

The initial agenda called *Specialist Agenda* (SPA) set by TACON is:

{ALL-RISE DATON MASON DATON CRITIC}

The preliminary design is done by ALL-RISE and DATON. Once the agenda has been set up by the *strategy level* knowledge module, TACON, the *monitor* takes the first module from it and executes it. All the modules in the agenda are executed sequentially till it is empty. All the sub-tasks of a module are also executed sequentially till the module is completed. TACON is activated only after the execution of CRITIC. Conceptually, TACON may be activated at any stage in case a more flexible mechanism is required.

4.3.5 Interactions between the knowledge modules

Figure 4.4 is the levels on the blackboard on which the CRITIC module posts and retrieves information from. Arrows indicate posting some information while the circles indicate retrieval of information.

4.4 Limitations of the DESTINY model

All the limitations of the DESTINY model, identified by this study, centre around the following fact:

> *there is no provision for handling the non-monotonic nature of structural design.*

This limitation leads to the criticism that apart from the *non-monotonicity* in structural design, there is hardly any point in applying artificial intelligence techniques because the other aspects of structural design can be very easily tackled by the conventional computing methods. It has been suggested that the architecture chosen for DESTINY, i.e., the blackboard architecture, is reckoned to be quite appropriate for handling *non-monotonic* situations. However, the DESTINY model does not appear to make proper use of this architecture. The execution of the different knowledge modules is *sequential*, which does not justify the use of a blackboard architecture nor does it tackle the *non-monotonicity* encountered so often in structural design. Because of these limitations, the DESTINY model suffers from two main drawbacks:

1. the lack of *knowledge* to tackle *non-monotonic* situations; and
2. the lack of proper *problem-solving methods* to handle such situations.

These points will be further discussed in Section 4.6, where the differences between the DESTINY and INDEX models will be presented.

4.5 The INDEX Model

4.5.1 Architecture

The architecture of the INDEX model is the same as the DESTINY model, i.e., the blackboard architecture. The reasons for selecting this architecture are the same as

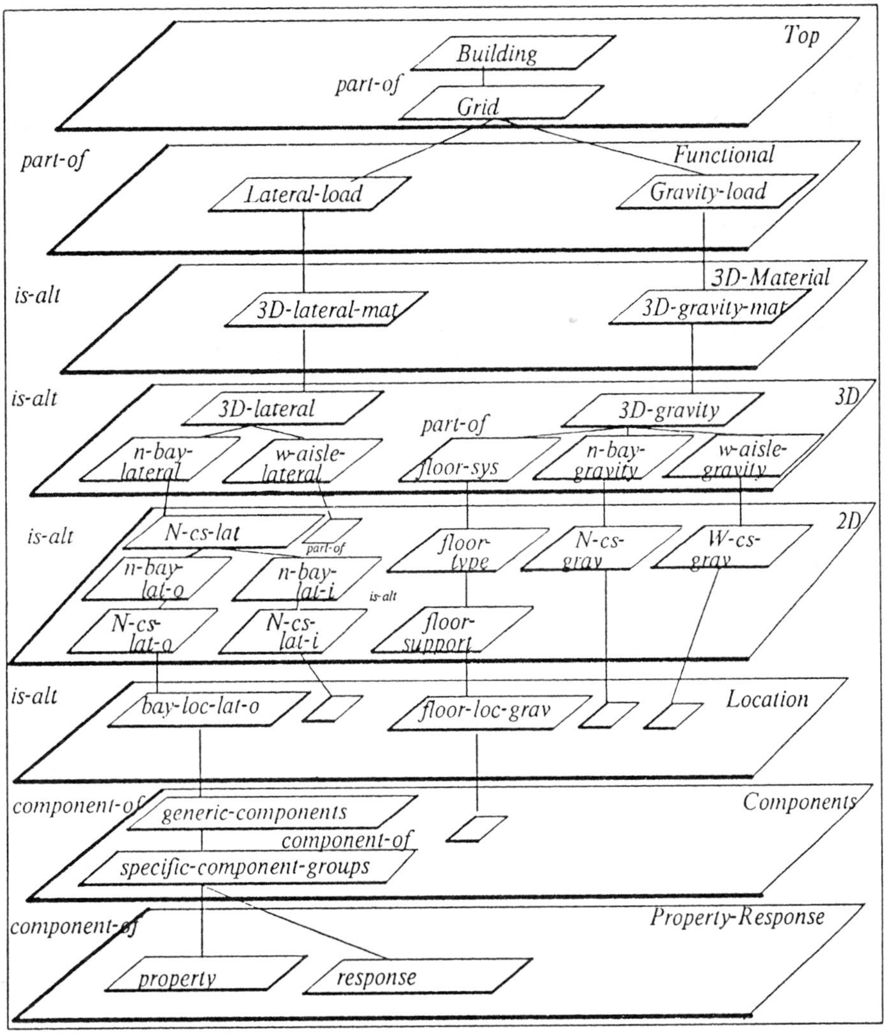

Figure 4.3 Abstraction hierarchy levels on the blackboard of DESTINY (taken from (1))

those for DESTINY.

4.5.2 Blackboard

The abstraction hierarchy on the blackboard of INDEX is similar to that on the blackboard of DESTINY in nature as the abstraction hierarchy is essentially the translation of different steps in the structural design process which is the same in both cases. This is why a separate abstraction hierarchy is not being given for INDEX.

4.5.3 Knowledge-Base

4.5.3.1 Brief description

The knowledge base of INDEX consists of a number of knowledge modules as shown in Figure 4.5. The knowledge modules are organised into a hierarchy of two levels, the specialist level and the resource level. The knowledge modules at the specialist level consist mainly of heuristics and other knowledge that are specialist-dependant. The knowledge modules at the resource level consist mainly of textbook knowledge. All the knowledge modules contain declarative as well as procedural knowledge. A brief description of the knowledge modules at the different levels is given below :

Specialist level : This consists of knowledge modules primarily containing experience-based heuristics. Of course, some textbook knowledge is also stored at this level.

This level consists of the following knowledge modules :

ALTSEL : This module is responsible for the ALTernative SELection of the feasible structural systems and deciding about different design parameters as the required frame spacing, whether to go for a single or a multi-bay system etc.

STRANEX : This module carries out the modelling and analysis of the chosen structural system by ALTSEL.

DETDEX : This carries out the detailed design, i.e. detailed proportioning and sizing of the components of the chosen structure.

EVALUATOR : This module evaluates the different alternatives generated by the system.

OPTEX : This module consists of various heuristics and rules to be used for the optimisation of the structures.

DESCON : This module is responsible for solving problems arising out of a change of specifications or constraints described in section 4.2.2. DESCON acts as a DESign CONsultant to the other modules in such situations.

Resource level : This level generally consists of algorithmic programs, e.g. structural analysis programs, standard codes, optimisation routines etc.

The knowledge modules at this level consist of the following:

STRANA : This module includes the STRuctural ANAlysis programs.

DETDES : This module is responsible for the DETailed DESign of the structure, i.e. detailed sizing of the components of the structure.

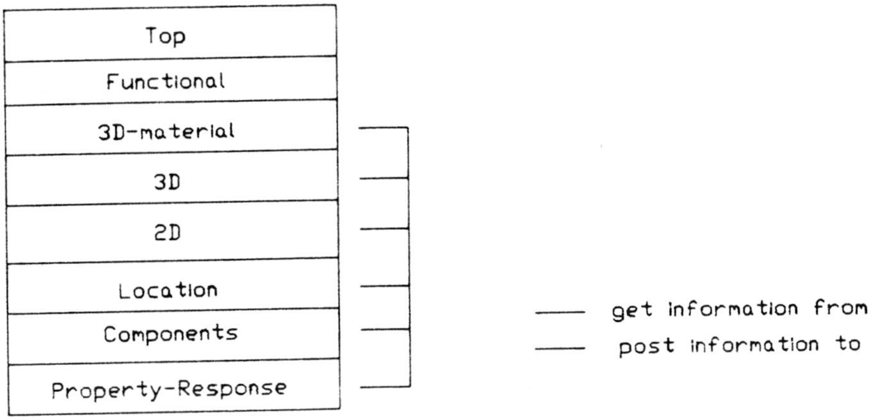

Figure 4.4 The blackboard levels CRITIC posts information on and gets information from (adapted from (1))

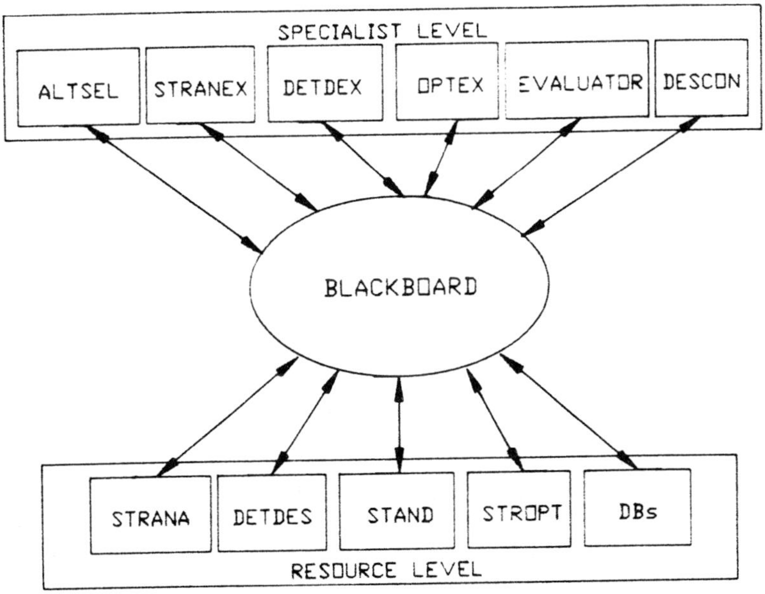

Figure 4.5 Schematic model of INDEX

STAND : This module includes the provisions of the applicable STANDards and is responsible for checking these standard constraints.

STOPT : This module consists of STructural OPTimisation routines.

DBs : These DataBases include the different dimensions and sectional properties of various structural sections, e.g. UBs, UCs etc.

4.5.3.2 Detailed description

This section describes the Specialist level knowledge modules in some detail.

ALTSEL

The main tasks before this module are as follows:

1. selection of the feasible alternative structural systems;
2. undertaking a preliminary analysis of the structural systems;
3. undertaking a preliminary sizing and proportioning of the components of the systems;
4. undertaking preliminary checks on the components;
5. undertaking posting relevant constraints for each of the alternative systems on the blackboard for a later use by the other modules; and
6. undertaking a preliminary evaluation of the systems.

The details of the different sub-modules of ALTSEL that perform the above-mentioned tasks and their implementation will be post-poned until Chapter 5.

STRANEX

The main tasks performed by the STRANEX module are as follows:

1. to model the structural systems generated by ALTSEL;
2. to select the structural analysis strategy, i.e., the appropriate type of analysis;
3. to prepare the input data for the analysis program; and
4. to return the output data from the analysis program.

STRANEX can be seen to be performing a similar task as SACON (6) discussed in Section 3.3.1. The different sub-modules of STRANEX are as follows:

MODELLER - models the structure for the analysis;

LOADEX - decides the type and magnitude of loadings imposed on the structure;

PLANNER - plans on the appropriate analysis program to be used; and

INTERFACE - prepares the data for the analysis program as well as receives the output data back from it.

The actual analysis is carried out by the analysis programs at the *resource level.* This knowledge module needs an interface with the analysis programs. An interface between PROLOG and FORTRAN77 was implemented for this purpose, details of which can be found in (7) and in Appendix I.

DETDEX

The tasks performed by DETDEX are as follows:

1. to size the different components of the structure;
2. to select the applicable provisions of the codes of practice; and
3. to check the constraints prescribed by the codes as well as other soft constraints;

The details of the different sub-modules of DETDEX will be postponed until Chapter 6. The functions of this module can be seen to be quite similar to that of SPEX (8) discussed briefly in Section 3.3.1. However, there are differences between SPEX and DETDEX that will become clear later in Chapter 6.

OPTEX

This module has the following tasks before it:

1. to formulate a model for optimisation of the structure;
2. to select the appropriate optimisation algorithm to be used;
3. to prepare the input data for the optimisation program; and
4. to get back the output from the optimisation program.

Following are the sub-modules of this module which perform the above-mentioned tasks:

MODELLER - formulates an optimisation model;

PLANNER - plans on the appropriate optimisation algorithm to be invoked;

INTERFACE - sends and gets back the data to and from the optimisation program.

This module also needs an interface with the optimisation programs. The same interface between PROLOG and FORTRAN77 discussed in Appendix I is used for the purpose as well.

DESCON

This module's function is to propose a solution to a design or partial design which is not satisfactory due to a change of specification or violation of some constraints. Thus, if a design is *unsatisfactory*, the following two possibilities exist:

a. the design is either *modifiable*; or
b. the design is not modifiable and a *re-design* has to be undertaken.

In case of a modifiable design, again the following two possibilities exist depending upon the extent and nature of modification to be carried out:

1. the modification will require a re-analysis of the structure; or
2. the modification will not require a re-analysis of the structure.

Based on the above criteria, the task of this module is to decide if a design is one of the following:

1. modifiable; or

2. re-designable.

This module is similar to the CRITIC module of DESTINY. However, there are very significant differences in their scope and operation which will be discussed in Section 4.5.

Once a design or partial design is categorised as discussed above, DESCON's tasks also include the following:

1. to formulate the revised set of constraints in cases of a *modifiable* design;
2. to suggest the exact nature of modifications to be carried out; and
3. to post constraints to the appropriate modules.

The details of the sub-modules of this module as well as their implementation will be given in Chapter 7.

The need for this additional module at the Specialist level will be discussed in Section 4.6. This module comprises the most important difference between the DESTINY and INDEX models.

EVALUATOR

Once all the alternatives generated by ALTSEL have been designed *satisfactorily*, all of them are passed to this module. The task before this module is to evaluate all the designs based on different criteria and rank them accordingly. This module was not developed in this work.

4.5.4 Inference Mechanism

Unlike DESTINY, the inference mechanism of INDEX is handled in two ways. Firstly, in routine situations, the sequence of execution of the knowledge modules is pre-defined, which is as follows:

(ALTSEL->STRANEX->DETEX->OPTEX->EVALUATOR)

Clearly, this sequence does not include the DESCON module. The reason is that DESCON may not be invoked in every case. Depending on the outcome of the other modules, DESCON may or may not be invoked. In case DESCON does have to be invoked, the control mechanism rests mostly in the hands of DESCON itself. In such cases, DESCON sets up the sequence of execution of any other module, if need be. The different cases that may arise before DESCON is invoked will be discussed in Section 4.6.2.

4.6 Comparison between the DESTINY and INDEX models

Although the INDEX model is based on the DESTINY model, there are significant differences between the two. The INDEX model can be seen as an extension of the DESTINY model, as will become clear later.

Figure 4.5 indicates that the INDEX model has two levels, viz., the specialist and the resource level. For the sake of uniformity, the terminologies used are the same as the DESTINY model (Figure 4.2). However, the DESTINY model proposes an additional level called the Strategy level. INDEX does not recognise the need for this level. The reason is that in DESTINY, the execution of modules is *sequential* which does not require a separate set of rules to define. The sequence may be pre-defined, which is how it is undertaken in INDEX. However, if there is a requirement to change the sequence, there will be a need for a set of rules. This will become clear from the discussions in Section 4.6.2. Apart from this difference, the other major difference is that of an additional module at the specialist level, DESCON. (In fact, there are two additional modules at the specialist level, viz., OPTEX and DESCON. However, the addition of OPTEX is not very important from the conceptual point of view of design. It can only be considered to be an additional facility of the system). This difference is an important one as the functions of this module are considered to be vital for an integrated design system. The rest of the modules at the specialist level can be seen to be quite similar in both cases.

4.6.1 Differences in the knowledge base

It was decided that an additional module is required to tackle some (if not all) problems detected by the CRITIC module of DESTINY. The detection of a problem (e.g., unconformity to standard requirements) is done by the resepective modules themselves in INDEX. The DESTINY model proposes passing back the control to either MASON or DATON. Unlike DESTINY, it is proposed to transfer the control from whichever module the problem is detected in to DESCON in every case[1] which, in turn, transfers the control to one of the other three design modules at the specialist level, viz., ALTSEL, STRANEX or DETDEX. In some cases, DESCON is proposed to take care of situations that may not have clearly defined constraints in structural engineering terms. For example, at some stage, a purlin (say) may have to be removed after it has been designed. The removal of a purlin is not an engineering constraint but may give rise to a lot of them. The input to DESCON in such cases is only the fact that the purlin has to be removed and not that the constraints relating to the stability of the rafters are violated. In these circumstances, DESCON's task is to infer the effects of any such changes to the already existing design or partial design and formulate new constraints and propagate them to the appropriate modules. CRITIC can only detect the problem and suggest whether the structure is:

1. unsatisfactory; or
2. modifiable; or
3. fixable; or
4. satisfactory.

The different cases when these values are assigned to a design are mentioned in the sub-section 4.3.3. However, the DESTINY model does not indicate whether CRITIC is also responsible for suggesting to the appropriate module the precise modifications to be carried out to a design. It is being proposed that very specialised knowledge may be required to accomplish such a task and, hence, the additional module, DESCON was developed. The basis on which CRITIC suggests whether a design is *modifiable* or *fixable* is whether or not a re-analysis of the structure is required. The only basis used by CRITIC to decide that is the amount of difference between the assumed and computed properties of the structure. In fact, such a decision could be quite a subjective one and may require other considerations to be taken into account. Furthermore, CRITIC cannot handle the emergence of new constraints or a change in

[1] The only exception is the *satisfactory* design in which case the design terminates at EVALUATOR.

specifications. CRITIC's scope is, thus, limited, and suffers from an important and serious drawback. DESCON of INDEX may be seen as an extension of CRITIC in that it can also handle situations where some constraint suddenly emerges as a consequence of either:

(i) been overlooked or ignored earlier; or

(ii) abrupt changes requested by the client; or

(iii) poor co-ordination between the structural and some other concerned designer (e.g., services).

DESCON's task is to suggest a *qualitative solution*. Subsequently, the actual *quantitative solution* is carried out by whichever module DESCON passes the control on to. DESCON is reckoned to be different from the rest of the specialist modules because its task is to suggest changes to an existing or a proposed design rather than designing a structure from scratch given the specifications. Its task is considered to be more difficult because it has to explore in a more *intelligent* way the alternative solutions that will force least modifications to the existing or proposed design at the least expense. The problem is more critical in case of an existing structure or a proposed design already fabricated. The most knowledge-intensive part of such an exercise is to find out the design *entities* that play the most important role in a particular case. The details of the problem-solving strategies utilised by DESCON will be discussed in detail in Chapter 7.

4.6.2 Differences in the interactions between the different modules

Figure 4.4 shows the levels on the blackboard of DESTINY where CRITIC posts information on. It is clear from this figure that in no case does it pass control back to the level where the feasible structural systems are selected. This is one improvement that the INDEX model proposes. In case of a *re-designable* design, the control is passed back to the ALTSEL module and a completely new alternative picked up generated earlier by it.

Figures 4.6a and 4.6b are diagrammatic representations of the interactions between different knowledge modules of INDEX and DESTINY repectively. Yet another major difference between the two models is illustrated by these two figures. The interactions between the modules of INDEX will be explained in some detail before highlighting the differences from that in the DESTINY model.

By considering Figure 4.6a, it is quite clear that the knowledge modules, ALTSEL, STRANEX, DETDEX, OPTEX and EVALUATOR are executed sequentially in a routine case. DESCON may or may not be invoked at all in particular cases. DESCON is only invoked when there is a problem detected by any module. The transfer of control to DESCON can take place at any stage of design apart from the routine case when the complete designs are passed to EVALUATOR to evaluate. Whenever a new constraint arises or there is a sudden change of specifications, the control is passed to DESCON. The transfer of control from DESCON to the other modules depends on the nature of the problem detected and the consequent decision taken by DESCON. The most straightforward case handled by DESCON is the *re-designable* design, in which case it simply passes control over to the ALTSEL module. The most complex case handled by DESCON is the *modifiable* design. In this case, the precise nature of modifiability is the main problem before DESCON. As far as the transfer of control is concerned, it is always to DETDEX. However, depending on whether or not a re-analysis is required DETDEX has to pass the control on to either STRANEX or to EVALUATOR after carrying out standards checking. For example, if it is suggested by DESCON that the design is *modifiable* and that the exact nature of

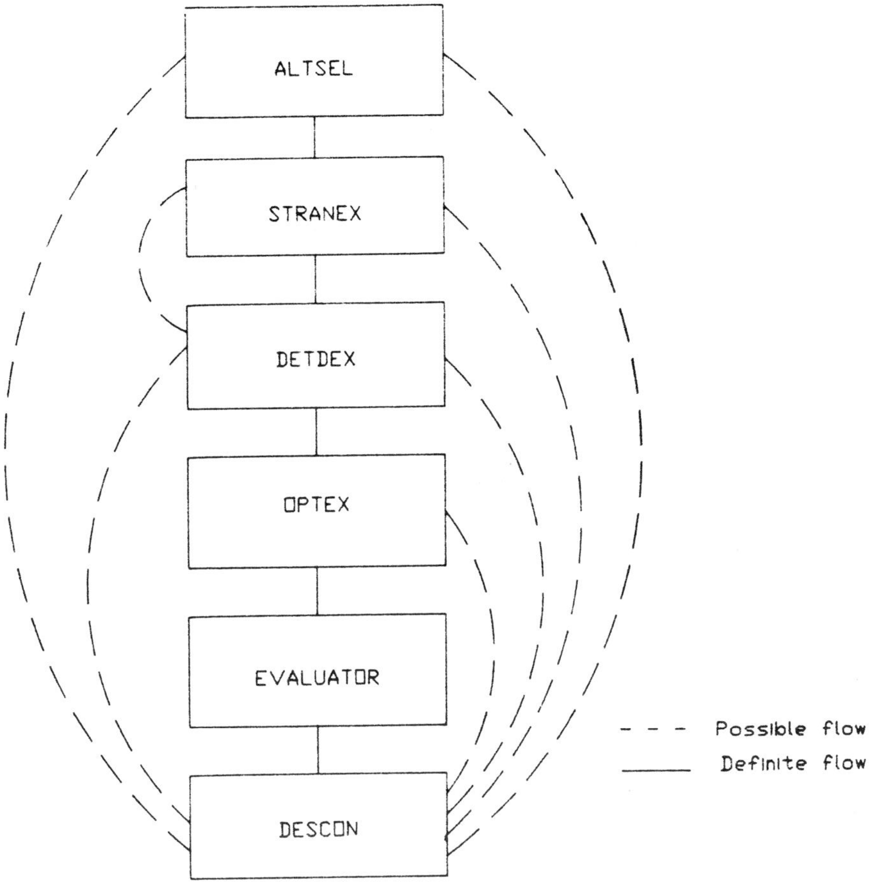

Figure 4.6(a) Interactions between the KMs in the INDEX model

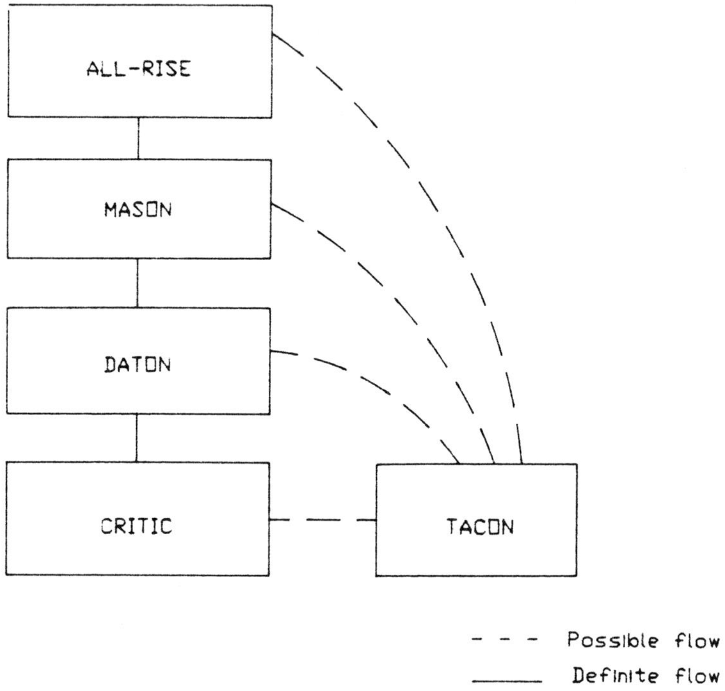

Figure 4.6 (b) Interactions between the KMs in the DESTINY model

modification to be carried out is to increase the section size of one of the members of the structure, DETDEX will pass control to STRANEX to carry out a re-analysis. Of course, even if the control is passed back to STRANEX, the standards provisions have still to be checked and then the design passed to EVALUATOR eventually. This process continues until a *satisfactory* is arrived at. So, every successful design has to terminate at EVALUATOR.

In DESTINY, the elements of the Specialist Agenda (SPA) are set by the Strategy level knowledge module, TACON. The TACON module is activated only after the execution of the CRITIC module. Although there is a mention of the possibility of activating TACON after the execution of each specialist level module, the mechanism is not very clear. In the INDEX model, it is proposed that there must be a facility in the model to take care of any *non-monotonicity* whenever it arises. It is thought that such a situation may arise at any stage and even partial designs may have to be assessed by DESCON. DESTINY quite obviously lacks this important feature.

In INDEX, part of DESCON's purpose can be seen to be quite similar to that of TACON's in DESTINY, i.e., invoking the specialist KMs. As far as invocation of DESCON itself is concerned, it will still be seen to be quite similar to that of TACON as well, i.e., every time any module posts information on the blackboard indicating a problem. But, the difference lies in the scope and purpose of knowledge in DESCON.

As discussed earlier, DESCON can be invoked in the following situations:

1. whenever a change of specifications or constraints occurs. This can happen in the following situations:

    a. at any stage in the design process, in which case even a *partial design* may be passed to it; or
    b. after the design has been completed, in which case the complete design will be passed to it.

2. whenever a violation of constraints occur.

The transfer of control to DESCON can be undertaken by any of the specialist KMs. In contrast, CRITIC module is only activated after the other KMs of the SPA have been executed in the sequence prescribed by the SPA.

The scope and purpose of knowledge inside DESCON has already been explained earlier, which is quite evidently an extension over the CRITIC module of the DESTINY model.

4.7 Summary and conclusions

A model for integrated structural design, INDEX, was presented. The model presented was an extension of an earlier model, DESTINY. Some limitations of the DESTINY model were pointed out and the INDEX model was the result of incorporating these lacking features of the DESTINY model. It was concluded that the structural design process was a *non-monotonic* process and any system developed for structural design needed to have facilities to handle this *non-monotonocity*. All the suggested extensions to the DESTINY model centre on this feature of design which is held to be fundamental.

References

1. **Sriram, D.**, "*Knowledge-based Approaches to Structural Design*", Ph.D. Dissertation, Department of Civil Engineering, Carnegie-Mellon University, Pittsburgh, U.S.A., 1986.

2. **Bell, T. and Plank, R.J.**, "*Microcomputers in Civil Engineering*, Construction Press, London and New York, pp.154, 1985.

3. **McDermott, D. and Doyle, J.**, "*Non-monotonic Logic I*", Artificial Intelligence, Vol. 13, 1980.

4. **Reiter, R.**, "*On Reasoning by Default*", TINLAP-2, University of Illinois at Urbana-Champaign,pp. 210-218, 1978.

5. **Hayes-Roth, Barbara**, "*A Blackboard Architecture for Control*", Artificial Intelligence, Vol. 26, pp. 251-321, 1985.

6. **Bennett, J.S. et. al.**, "*SACON: A Knowledge-based Consultant for Structural Analysis*", Technical Report STAN-CS-78-699, Stanford University, Palo Alto, California, U.S.A., 1978.

7. **Kumar, B., Chung, P.W.H. and Topping, B.H.V.**, "*Approaches to FORTRAN-PROLOG interfacing for an Expert System Environment*" in The Application of Artificial Intelligence Techniques to Civil and Structural Engineering, Ed., Topping, B.H.V., Civil-Comp Press, Edinburgh, pp. 15-20, 1987.

8. **Garrett, J.H., Jr. and Fenves, S.J.**, "*A Knowledge-based Standards Processor for Structural Component Design,*" Report No. R-86-157, Department of Civil Engineering, Carnegie-Mellon University, Pittsburgh, U.S.A., 1986.

Chapter 5

ALTSEL: The Preliminary Design Module of INDEX

5.1 Introduction

The overall INDEX model has been described in Chapter 4. This chapter describes the preliminary design module of INDEX, ALTSEL. ALTSEL represents a simple but effective rule-based prototype for the preliminary design of industrial buildings. The use of rule-based programming is illustrated by describing the development of ALTSEL. A description of knowledge elicitation techniques and some practical lessons learnt from using them will also be discussed. Some useful features of the Edinburgh Prolog Blackboard Shell, which simplified the development of ALTSEL, will also be discussed which shed light on the utility of blackboard architecture for a knowledge-based design system.

5.2 Some major features and components of ALTSEL

The architecture of ALTSEL is a blackboard system. It consists of different knowledge-modules surrounding and communicating through the blackboard as discussed in Section 2.7.2.2. The input to ALTSEL is the general layout and other spatial constraints of the building. Since generally the layout of the design is fixed by architectural design, the domain of the system is restricted to structural design.

5.2.1 Blackboard

As already described in Section 4.5, the general descriptions of the blackboard of INDEX apply to that of ALTSEL, too. In fact, all the modules of INDEX can be seen to be separate knowledge-based prototypes. ALTSEL's blackboard is divided into different parts which contain entries posted on it by the different sub-modules of ALTSEL in the course of the solution process. The levels on the blackboard of ALTSEL may be seen as a hierarchial decomposition of the preliminary industrial building design process.

The abstraction hierarchy on the blackboard of ALTSEL is given in Figure 5.1. This abstraction is given in terms of 'specifications' and 'products'. This abstraction may be seen as a hierarchial decomposition of the design process as the solution emerges on the blackboard. Every level in this figure is a 'product' of the level above it and forms the 'specifications' for the one below it. This fact is represented by the 'prod-of' and 'spec-for' links between the different levels. This figure does not include the whole design process and is confined to the preliminary design undertaken by the ALTSEL module.

5.2.2 Knowledge Base Development

5.2.2.1 Knowledge Elicitation

Introduction

The construction of a knowledge-based expert system is an attempt to embody the knowledge of a particular expert within a computer program. The knowledge used in solving problems must be elicited from the expert to incorporate in the expert system. It is recognised that the elicitation of knowledge from the experts is one of the major

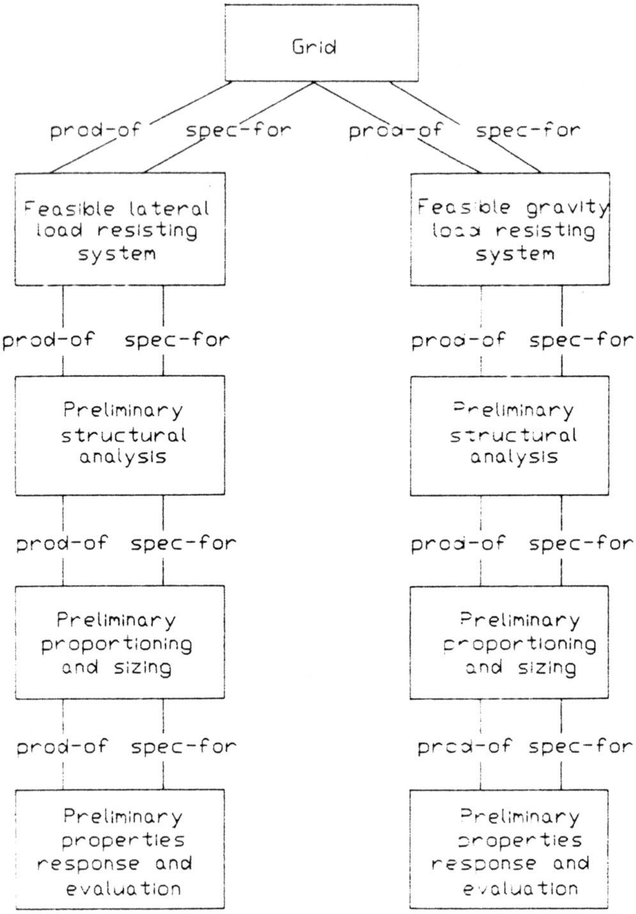

Figure 5.1 An Abstraction hierarchy on the blackboard for the solutions generated by ALTSEL

obstacles in the construction of expert systems. In many cases, the main reason for this is that experts find it hard to articulate and make explicit the knowledge they possess and use. An important part of a knowledge enginner's job is to help the expert to structure the domain knowledge and to identify and formalize the domain concepts. Although a number of knowledge elicitation methods do exist (1), the area is not well understood and few tools exist to mechanise the process.

In the following sections, a simple model of knowledge elicitation will first be presented followed by some specific techniques. Finally, in Section 5.2.2.1, an account of the knowledge elicitation process for ALTSEL will be presented to highlight some of the practical issues.

A Framework For Knowledge Elicitation

The framework is based on three generally accepted ideas:

1. there are different types of knowledge;
2. there are different knowledge elicitation methods for different types of knowledge; and
3. the knowledge elicitation process can be divided into sub stages.

There is no doubt that there are different types of knowledge, even in a single domain of expertise. However, it is not clear how knowledge should be classified into different types. "Finding a way to taxonomise knowledge on a principled basis is a difficult and ambitious task that has eluded philosophers for thousands of years" (2). For the practical purpose of building expert systems, knowledge may be conveniently divided into three types: *facts, conceptual structures* and *rules*. *Facts* are simply a glossary of terms and a list of domain entities. In an engineering domain, this type of knowledge may be a collection of engineering concepts and the names of the components of a particular structure or any other engineering artifact. The second type of knowledge, *conceptual structures*, describes the relationships between identified concepts and components. Finally, *rules* are the reasoning part of the domain knowledge. Facts and conceptual structures are reasonably static and are easier to elicit than rules. Figure 5.2 illustrates a simple but a natural sequence of knowledge elicitation process.

In each part of the cycle, a suitable elicitation technique should be used. Some studies have been carried out to match techniques with types of knowledge (1,2). In the next section, for each type of knowledge, a knowledge elicitation technique that has been identified as particularly suitable, is described.

Techniques

There are two classes of techniques for knowledge elicitation as follows:

(1) **psychological technique**, which involves some kind of interaction between the knowledge engineer (KE) and the domain expert (DE); and
(2) **machine induction**, in which the computer induces rules from examples automatically.

For a domain such as structural design, *machine induction* seems inappropriate. Bloomfield (3) developed a set of criteria for selecting domains suitable for the elicitation of knowledge by *machine induction*. One such criterion is that "any chosen

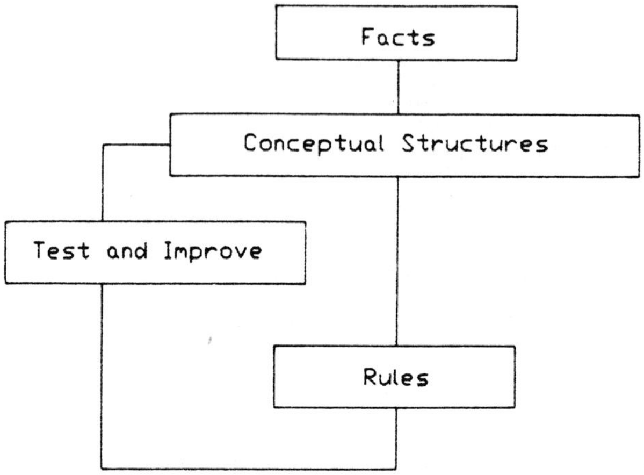

Figure 5.2 Sequence of Elicitation Process

domain must contain sufficient examples that it is possible to construct a training set which constitutes a comprehensive encapsulation of expertise in that domain". Structural design expertise cannot be completely encapsulated in examples. Hence, only *psychological techniques* are considered.

Interviews

Direct interviewing is the technique most familiar to KEs and DEs. It is considered good practice to start the knowledge elicitation process using a technique that the DE feels comfortable with. An interview may range from an informal chat to a highly structured discussion. Some interesting openers for an interview may be:

- if you had a good new graduate just starting to work for you what would you expect him to have learnt after six months ?
- you find a book on your application area which turns out to be the book you wish you had started in the field. What chapter headings are in it ?

Using this technique, a lot of information about the terminology and the main components of the domain can be generated in a relatively straightforward way. The problem is how to probe further so that ideas may be pursued to a greater depth. To ensure that an interview is productive, the KE should have a good questionnaire prepared beforehand to help him direct the discussion. Instead of just open questions he needs to have some clear and specific ones. The DE may also be asked to prepare and deliver an introductory lecture.

Concept Sorting

Experts use specialist knowledge to solve problems; they are likely to have a global perspective on how a domain is organised. Concept sorting is appropriate where there is a large set of concepts which need to be organised into a manageable form. The basic procedure is similar to the categorical knowledge elicitation technique described by Regan (4):

1. collect a set of concepts in the domain. This can be obtained from the literature or from an introductory talk or from the DE;
2. write each concept on a small card;
3. ask DE to sort the cards into groups;
4. ask DE to label each group;
5. discuss with DE about each group to find out its characteristics;
6. ask DE to specify the relationship between the groups and to organise them into a hierarchy.

Protocol analysis

In this technique, the DE's behaviour is recorded (either video or audio) as they work through a problem or task, and the protocol is transcribed and analysed. In this way, the KE is not only given the answer to the problem but also the information about the problem solving process itself. In practice this technique is found to be very helpful. Though DEs may have difficulty in stating the general rules that they use, they can usually identify the specific rules that they are applying. However, it is easy for familiar ideas to be taken for granted, so they need to be kept aware of any tendencies towards omitting *trivial* details. For this technique to be effective a representative set of problems has to be chosen, otherwise there could be serious errors of omission.

There are three different ways of generating protocols:

- think-aloud protocols - the DE thinks aloud during the solving of a problem;
- retrospective verbalization - the DE completely solves a problem before reporting how it was solved;
- discussion protocols - a small number of DEs discuss with one another as they attempt to solve a problem.

Each of these variations has its own advantages and disadvantages. An important problem with think-aloud protocols is that the reporting may interfere with the DE's task performance. Related to this is any need to conform to real time constraints. For example, solving a mathematics problem allows the mathematician to stop and ponder. However, an operator dealing with an emergency may require immediate responses. These criteria may help when having to decide between think-aloud protocols and retrospective verbalization.

Expert system projects are often based on collaboration with a single DE. In fact most of the literature recommends this (5). However, discussion protocols are helpful because they provide different perspectives on how a problem may be solved by clarifying alternatives and resolving conflicts. The problem here is that of managing the discussion. Avoiding the problem, the strategy that Mittal and Dym (6) adopted was to interview one DE at a time. Although this technique worked for them, it provides very little opportunity for the DEs to interact with one another and to discuss issues.

A potentially useful computer tool for collaborative problem-solving in face-to-face meetings is Colab, which has been created at Xerox Parc (7). This project advocates the use of computers rather than a passive medium like chalkboards in meetings. The idea is that in the meeting room each person has a keyboard and mouse on his table and there is a very large screen in the front of the room. Each person may retrieve information from the computer and can easily write and draw to the screen by using the keyboard and mouse in front of him. In this mode of working a meeting may be dynamic and interactive, and at the same time all the text and sketches that have been generated in the meeting are automatically stored on the computer. The abundance of information is conveniently accessible for analysis when needed.

Rapid Prototyping

The most obvious technique for testing and improving an expert system is rapid prototyping. The DE is confronted with the behaviour of an unfinished version of the system which is modified in the light of his/her comments. Each iteration brings the behaviour of the system closer to completion although, since it is often carried out without a clearly defined notion of completion, it is perhaps better thought of as iteration towards adequate achievement.

Summary of the Techniques

These are just some of the techniques that have been identified as useful. They should be viewed as complementary rather than competitive with one another because different techniques may be used to capture different types of knowledge more effectively. Interviews are good for gaining an overall view of the domain; concept sorting is good for structuring the domain; and protocol analysis is good for collecting rules. The main point is that the KE needs to be aware that there are different techniques that may be applied. Their usefulness also depends very much on the individual KEs. Factors such as KE's knowledge of the problem domain and how well they get on with the DE matter a lot.

From the description of different techniques, it should also be clear that feedback plays a very important role in knowledge elicitation. It is highly unlikely that a DE can impart all relevant knowledge at one meeting even if the domain is extremely simple. The question is then *What form of feedback should be provided?* An obvious but important comment is that what is fed back should be familiar to the DE so it may easily be understood and commented upon.

Knowledge Elicitation for ALTSEL

The following sections describe the experiences gained in the knowledge elicitation process undertaken for ALTSEL. The relevance of the different techniques described earlier will become evident in these sections.

Meeting the Experts

The KE contacted a consultancy company which specializes in designing industrial buildings. Four meetings took place, with each lasting approximately three hours. The following is a short commentary on what happened during each of these four meetings.

First Meeting

At this meeting the KE met the DE, a design engineer with many years of experience. The DE knew that he had expertise and was sceptical that a computer could perform the same function as himself. So, throughout this meeting, the KE tried to convince the DE by describing to him how expert systems work and showing him the listing and runs of a simple prototype design checker (see Appendix II) developed in the earlier part of this work. The DE remained unconvinced. He had two basic doubts:

1. How could a computer reason except through obeying instructions?
2. Every design is different; how could a single set of rules apply to all designs?

The KE left the meeting frustrated and discouraged. Nonetheless, they agreed to have a second meeting two weeks later.

Second Meeting

At this meeting, there were three DEs: the previous design engineer, another design engineer and an expert in computer aided design. The first part of the meeting was very much the same as the previous one with the KE trying to convince the DEs that expert system technology was workable.

However, this time the KE had a copy of a diagram with him that illustrated an abstraction hierarchy of the design of a building. The diagram (Figure 5.3) is a simplified version of another diagram that the KE had found in the literature. The original abstraction diagram was developed by Sriram (8) for his work on a knowledge-based system for designing buildings. He showed this to the DEs who immediately identified that this reflected how they carried out design. In other words, the diagram helped the DEs conceptualise their own thinking processes and relate them to those of an expert system.

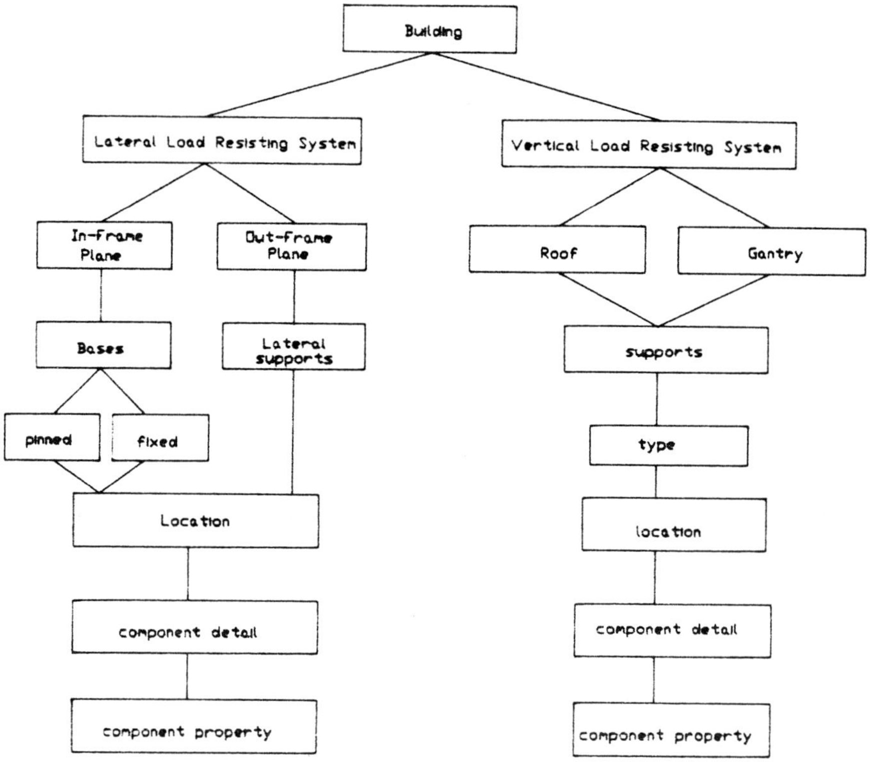

Figure 5.3 A simplified abstraction hierarchy of industrial building design process

Some time later the KE was left with the second design engineer to work through a design problem that he had recently solved. The DE was quite happy to explain how he had made certain decisions when he was asked the question "*Why?*". The DE also pointed out some literature that practicing engineers read.

From the informal protocol collected, the KE was able to produce ten rules. Further, the KE was able to identify and glean more rules from the literature that he had read earlier. The KE then built a prototype that took a specification as input and produced alternative feasible structural systems as output, alongwith a recommendation of which one of these alternatives was most favourable for further analysis and detailed design.

Third Meeting

This meeting took place a month after the previous one. When the DE saw the runs and rules of the system, he was very surprised by the progress that had been made. He spent most of the time in this session commenting on the rules.

After this session the KE was able to refine his rule-set and try the system on other problems that he had collected from literature.

Fourth Meeting

At this meeting the DE introduced three new problems and described to the KE how he had solved them.

To date, the knowledge-base has over a hundred rules. The table in Figure 5.4 gives a breakdown of the sources of the rules.

Discussion and Conclusion

The most important lessons learnt from this project concerned the following:

1. KE's familiarity with the domain; and

2. KE's initial approach.

It is concluded that the factor that seemed to play the most important role in particularly speeding up the knowledge elicitation process was the KE's familiarity with the domain. It is a well known fact these days that computer scientists claim to be as equally effective knowledge engineers as someone from the domain. A very common recommendation is that any person can be a good knowledge engineer after spending sometime in a consultant's office. The experience of this project is in total disagreement with this view. There were a number of occasions when the KE helped the DEs articulate their ideas. The only factor seemingly helping the KE in doing so was his familiarity with the domain. That sort of familiarity may not be acquired in a week's time in an office.

Another useful feature that seemed to help the whole process was the abstraction hierarchy of the design process. Although a KE might not always be able to generate a relevant diagram by himself, he should be able to produce one with the assistance of the DE. The *concept sorting* procedure (described in Section 5.2.2.1) is a good bottom-up technique to use. Any diagram during the knowledge elicitation phase can form useful documentation of the system.

Protocol analysis, or more precisely, studying case histories, was found to be a very useful way of generating rules. However, it is interesting to note that only a third of the rules were elicited directly from the DE (see Figure 5.3). Following through the

Source	Number of rules
Design Engineer	35
Literature	53
Other Sources	22

Figure 5.4 Breakdown of rules according to their sources

information provided by the DE as to where to look for further or more detailed information yielded much dividend.

The KE found decomposing the problem, especially the preliminary design part into sub-problems at an early stage was an extremely important step in formalising the domain knowledge. Once the problem was decomposed it not only helped the DE to recall and provide the relevant pieces of information, it also helped the KE to pick out relevant material from other sources. From the system construction point of view it was also very helpful because the knowledge base could then be divided into smaller modules making them easier to maintain.

The DE was surprised and impressed by the result of prototyping. It is definitely a very useful way for getting feedback from the DE. In this case it was a shame that due to geographical constraints the DE did not see the prototype running but could only comment on the output of the program.

When using prototyping as a technique to obtain feedback, the KE found it necessary to guard against letting the documentation slip. It is easy to get into the habit of making quick changes to the system without keeping a record of the changes made, thus making the system difficult to modify and maintain in the future.

At the time, it seemed quite obvious to approach the DEs by first convincing them about the potentials of knowledge-based expert systems. As pointed out earlier, quite a number of hours were wasted in doing so. The lesson, thus, learnt was that an attempt should be made to collect information from the DEs without making tall claims about the proposed system. It now seems important to stress on the fact that the proposed system is only intended to assist the DEs and not replace them. It seems important to reassure the designer's supreme role in the design process in relation to any computer program. This point might sound trivial but certainly appears to be a very important one.

5.2.2.2 Knowledge Representation

Introduction

The knowledge representation formalism used in ALTSEL is the same as that for INDEX and that is as *production rules*. The knowledge-base of ALTSEL is organised into different sub-modules as shown in Figure 5.5. As discussed in Section 4.5.3, ALTSEL itself is at the *Specialist level* of INDEX and thus, almost all the knowledge contained in ALTSEL is heuristics obtained from either experts or different literature (Figure 5.4). As already mentioned in Section 4.5.3.2, the purpose of ALTSEL is as follows:

1. selection of the feasible alternative structural systems;
2. carrying out a preliminary analysis of the structural systems;
3. carrying out a preliminary sizing and proportioning of the components of the systems;
4. carrying out preliminary checks on the components;
5. posting relevant constraints for each of the alternative systems on the blackboard for a later use by the other modules; and
6. carrying out a preliminary evaluation of the systems.

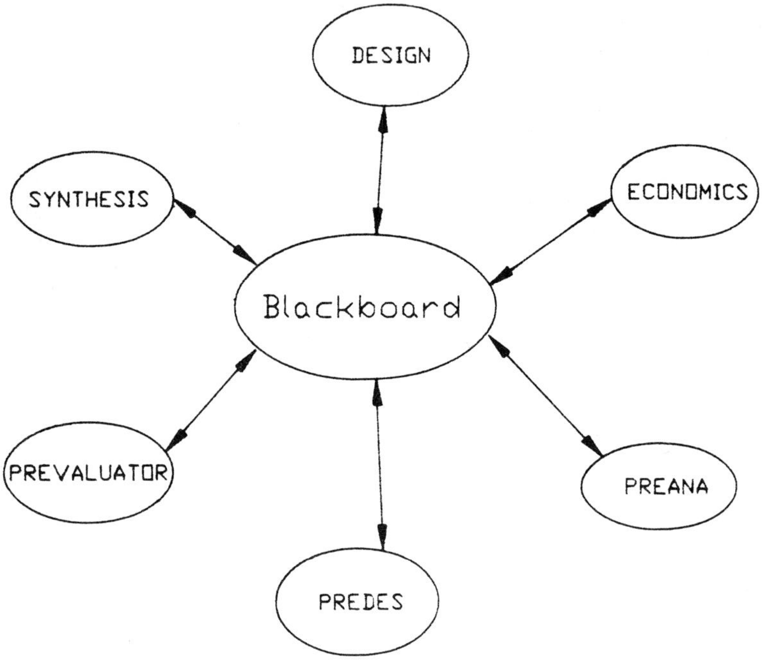

Figure 5.5 The sub-modules of ALTSEL

A later section describes the different knowledge modules of ALTSEL that perform these tasks.

Types of constraints

The identification and proper use of various constraints constitutes probably the most important aspect of any design. The different types of constraints considered by different knowledge modules of ALTSEL are different and depend upon the task being performed. The constraints considered by the sub-modules of the ALTSEL module are mostly *external*. For a comprehensive description of different types of constraints in structural design, reference can be made to (8). *External* constraints are the constraints that are not in the hands of the designer. In other words, these constraints are *external* to the designer. These constraints are mostly governed by the requirements of the client.

In order to take various decisions listed above, the various decisive factors were first to be found out. On the basis of discussions with practising engineers and a study of the design literature used by them to assist in taking these decisions, it was concluded that the following parameters played the most decisive roles in the preliminary design of industrial buildings:

1. span;
2. loads;
3. allowable pitch;
4. intended industrial process to be carried out in the building; and
5. any other client-related constraints.

Thus, all the rules in ALTSEL have one or more of the above-mentioned parameters as constraints to be satisfied. The constraints considered by the SYNTHESIS sub-module in deciding about the feasible lateral load resisting systems are shown in Figure 5.6. Figure 5.7 shows the considerations used by the PREVALUATOR sub-module in evaluating the different feasible systems.

Sub-modules of ALTSEL

SYNTHESIS - This module is responsible for selecting the feasible structural systems for the building in question. A typical rule from this rule is as follows:

```
if    [problem,span(X),true]
and    [problem,int_stanchion(no),true]
and    holds X > 60
then   true
to    add[lateral_load_sys,single_span_portal,true]
est   synthesis(1).
```

This rule is in the Edinburgh Prolog Blackboard Shell syntax and states that if the span of the building is more than sixty metres and that there are no internal columns allowed in the building then one alternative for feasible lateral load system is the single span portal frame. The different parts of the rule are explained in Section 2.7.2.3 However, the value of *est* in this rule is different from the default provided by the shell. This will be explained later in the Section 5.3.2.

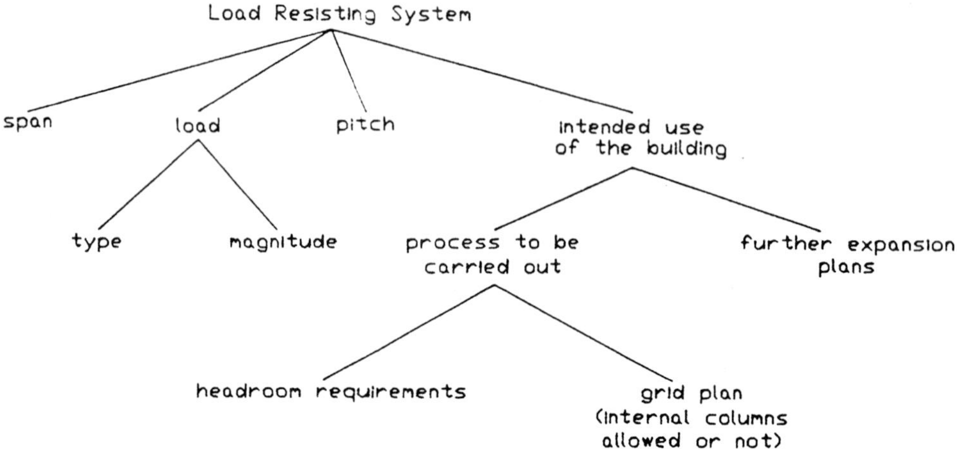

Figure 5.6 Constraints considered by SYNTHESIS

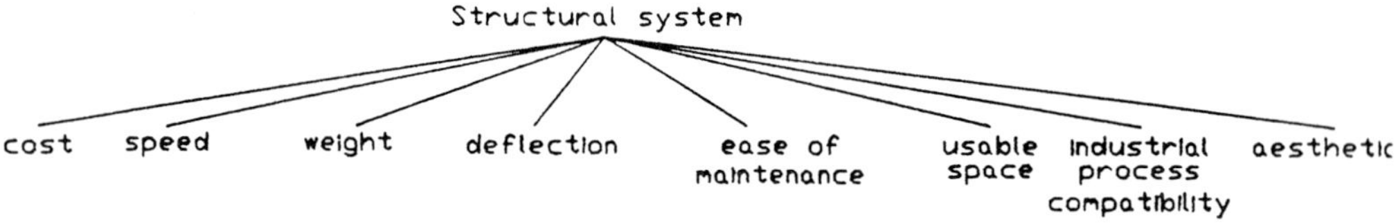

Figure 5.7 Evaluation characteristics considered by PREVALUATOR

In addition to selecting the feasible alternative structural systems, SYNTHESIS also decides about the frame spacing, appropriate systems for roof and the sides and their claddings.

PREANA - This module is responsible for undertaking the preliminary analysis of the alternative systems generated by SYNTHESIS. The rules in this sub-module are mostly analysis formulae for different types of structural systems. Also included in this sub-module are rules to decide what type of analysis should be undertaken. For example, it carries out a plastic analysis for a portal frame whereas for a truss it can only undertake a routine elastic analysis. Following is a typical rule from this sub-module which applies to single span portal frames whose bases are fixed:

```
if      [synthesis,lateral_load_sys(single_span_portal),true]
and     [problem,bases(fixed),true]
and     [problem,span(L),true]
and     [problem,load(W),true]
and     [problem,eaves_ht(H1),true]
and     [problem,pitch(Y),true]
then    moment1(H2,((L/2)*tan(Y)),X,((H1/(H1+H2))*(W*L^2)/16))
to      add[preana,sspfb_pla_mom(X),true]
est     preana(1).
```

The successful execution of this rule will add an entry on the blackboard under the index *preana* and the fact will be the value of the plastic moment for the single span portal frame alternative. The actual *moment* is calculated by the PROLOG clause 'moment1(H2,((L/2)*tan(Y)),X,((H1/(H1+H2))*(W*L^2)/16))' in the *consequent* (the 'then' part) of the rule. The other rules are quite similar in nature as the one quoted above as well as being of little research value and, thus, are not being quoted.

PREDES - PREDES is responsible for carrying out the sizing of the different members of the alternative structural systems generated by SYNTHESIS and whose preliminary analysis is already carried out by PREANA. The following is a rule from this module:

```
if      [preana,ssp_pla_mod(X),true]
and     [problem,ssp_rafters_sec_ext_cons(no),true]
and     [problem,ssp_stanchion_sec_ext_cons(no),true]
then    get_section(X,A,Y)
to      add[single_span_portal,ssp_feas_sec(A),true]
        and[single_span_portal,ssp_zp_provided(Y),true]
        and[single_span_portal,ssp_rafters_sec(ub),true]
        and[single_span_portal,ssp_stanchion_sec(ub),true]
est     predes(3).
```

This rule's execution finds out the feasible section from a database of Universal Beam and Column sections given the plastic modulus of the portal frame and also stating that there are no external constraints on the dimensions of either the stanchions or the rafters. Both the stanchions as well as the rafters are assumed to be of the same uniform sections.

ECONOMICS - This sub-module is fully based on heuristics obtained from the results of a research project on the comparative costs of single-storey steel structures (9). The firing of the rules of this sub-module depends on the presence of a particular type of lateral load system on the blackboard. The following rule illustrates this and also illustrates the type of knowledge contained in this sub-module:

```
if      [problem,span(X),true]
and     [synthesis,lateral_load_sys(roof_truss),true]
and     [problem,pitch(Y),true]
and     holds((13.3 =< X,X =< 26.7,Y > 0.3))
then    true
to      add [economics, lateral_load_sys_eco(roof_truss),true]
est     eco(1).
```

The rule simply states that if roof truss is one of the feasible systems and the span is between 13.3 and 26 metres and the pitch of the roof is greater than 0.3 radians then roof truss will be the most economical structural system.

DESIGN - This sub-module's purpose is not directly concerned with the preliminary design stage. However, it has a very important purpose to serve in so far as the whole design process is concerned. This sub-module finds out any relevant constraints that should be satisfied in the detailed design stage of any alternative structural system. The reason for keeping this sub-module in the ALTSEL module was that the knowledge contained in this sub-module is very much related to the feasible structural systems generated by SYNTHESIS. Also, all the constraints to be satisfied in the design of any structural system should be propagated to the other modules the moment a particular structural system is found to be feasible by SYNTHESIS. In some ways, DESIGN may be regarded as a *meta-module*, i.e., a module that operates above all the other modules. Following is a typical rule form this module:

```
if      [synthesis,lateral_load_sys(single_span_portal),true]
and     [problem,span(Y),true]
and     holds(Y > 10)
then    output_message('The following things should be considered in the
detailed design stage of single span portal alternative :-
        1.pitch should be kept low because greater slope will give rise to greater
spread at knees which can cause problems with cladding,
        2.horizontal thrusts should be carefully examined and the foundation
designed accordingly,
        3.haunch should be provided at the eaves and the ridge should be
deepened because the maximum bending moment will occur at the knees.')
to      add[design,design_cons(single_span_portal),true]
est     des(3).
```

This rule applies to the constraints for the single span portal frame alternative.

PREVALUATOR - This sub-module is responsible for carrying out a relative evaluation of the different feasible structural systems generated by SYNTHESIS. The different criteria considered by PREVALUATOR in doing so is given in Figure 5.7. The different criteria are given different weightage as suggested by different practicing engineers and a final *value* is given to each alternative structure. The best structure is the one with lowest *value*. This sub-module could not be fully developed due to problems in quantifying subjective considerations such as *aesthetics*.

5.2.3 Problem-solving Methods Used

All the different problem-solving strategies in Knowledge-Based Systems Technology adopt one of the following two approaches (10) :

1. Formation approach and

2. Derivation approach.

The *formation approach* involves the formation of the most appropriate solution by putting together the different components of a complete solution stored in the knowledge-base at different levels. In contrast, the *derivation approach* involves picking up the most appropriate solution from a set of pre-defined solutions already stored in the knowledge-base.

It is quite evident that the formation approach is probably a more general and intelligent way of solving a problem. However, for the domain we are working in, we found that the derivation approach provided an easier way of solving the problems. This was decided after experimenting with the formation approach. Hence, ALTSEL utilises the derivation approach to solving the problem. This is in contrast to HI-RISE which utilises a formation approach (10). Figure 5.8 is an inference network for the selection of feasible lateral load systems. The knowledge-bases of ALTSEL consist of different feasible solutions for different situations. In doing so, it proceeds ahead by handling different constraints (11), which consists of the following :

1. constraint formulation,

2. constraint satisfaction and

3. constraint posting.

This concept of constraint handling is accomplished in the system by first *satisfying the constraint* for a particular alternative, *looking for any constraint* associated with the alternative which will be used by other modules, i.e., *constraint formulation*, and *posting* it to the appropriate module which is supposed to use it later on, i.e, constraint posting. The *constraint satisfaction* operation is carried out by all the different sub-modules. The *constraint formulation* is done by the DESIGN sub-module as discussed in an earlier section. The *constraint posting* is also accomplished by the DESIGN sub-module but the way it works is by actually exploiting an inherent feature of communication between different *knowledge sources* of the *blackboard architecture*. This is an example of an important use of the *blackboard*. The following rules (one from the SYNTHESIS, one from the DESIGN and one from the standards processing sub-module, STAPRO) will illustrate this idea.

```
if      [problem,span(X),true]
and     [problem,int_stanch(no),true]
and     holds(X =< 60)
then    output_message('A single span portal would be  feasible.')
to      add[synthesis,lateral_load_sys(single_span_portal),true]
est     synthesis(5).
```

```
if      [synthesis,lateral_load_sys(single_span_portal),true]
then    true
to      add[design,single_span_portal(elastic_defl_check),true]
est     design(12).
```

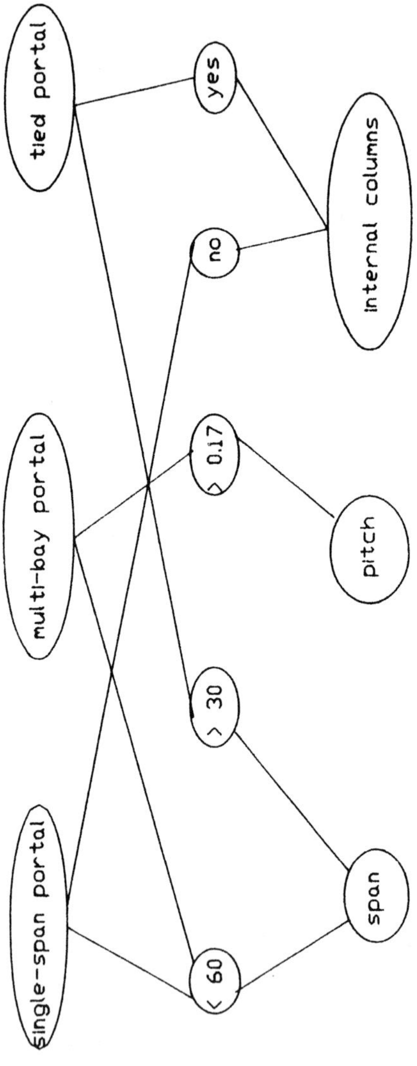

Figure 5.8 Inference network for lateral load resisting frames

```
if     [synthesis,lateral_load_sys(single_span_portal),true]
and    [design,single_span_portal(elastic_defl_check),true]
then   check_clause(sec. 5.1.2.3)
to     add[conformance(portal_frame,5.1.2.3,deflection),satisfied,true]
est    design_check(12).
```

where 'check_clause' is a PROLOG procedure that checks the provisions of Section 5.1.2.3 of BS5950.

In this example, the constraint has been posted on the blackboard with the index 'design' by the DESIGN sub-module. It may be accessed by any other module using this index. For example, the last rule from the standard provision checking module states that, if there is an entry on the blackboard that says that the deflection should be checked by elastic methods, then the provisions of clause 5.1.2.3 of the standard must be satisfied. It is quite evident how giving appropriate *indexes* to different *entries*, may be used in *partitioning* the *blackboard* which can be further used in *formulating* and *propagating* constraints to other modules.

5.2.4 Explanation facilities

Some explanation facilities have also been incorporated in ALTSEL. Basically, two approaches have been investigated. One was to have associated explanation with every conclusion the system may arrive at. The other approach adopted was to use the front-end facilities of the shell and generate explanations using them.

The explanations one may obtain from the system are :

1. the rule or set of rules that forced a particular conclusion;
2. the current entries on the blackboard;
3. the reasons for concluding something;
4. the set of rules that were successful at the end of a session; and
5. the details of any of the alternative feasible solutions generated by the system.

The examples given in Appendix III will illustrate these explanation facilities of the system.

5.3 Implementation

5.3.1 General Description

The current version of ALTSEL module incorporates over one hundred rules. It is implemented on a Sun 3/50 workstation. The system has knowledge about the following types of steel frames :

1. portal frames;
2. roof trusses and columns; and
3. beams and columns.

Apart from these, it also has rules for incorporating gantries for the design of gantry cranes if required. The solution search space for the feasible lateral load resisting systems is shown in Figure 5.9. The search strategy adopted is the breadth-first search. The system generates all the solutions at one level before going on to the next level. Figure 5.10 illustrates the search procedure adopted by the ALTSEL module. One drawback with this approach has been the lack of transparency of the system. The user does not get the complete details of a particular alternative at a glance. To overcome this drawback, the user is given the facility of obtaining the complete details of any alternative solution generated by ALTSEL at the end of the session using the 'show_details_of' command (see Appendix III).

The ALTSEL module has six sub-modules as shown in Figure 5.4, viz., SYNTHESIS, PREANA, PREDES, ECONOMICS, DESIGN AND PREVALUATOR. Based on the rules in these sub-modules, the system is able to select the feasible alternatives for the lateral load resisting main frames for the industrial building in question.

Although some of the rules are based on discussions with working design engineers, most of them are taken from published literature from various steel section and frame manufacturing and fabricating firms and organisations such as the Steel Construction Institute (formerly known as the Constructional Steel Research and Development Organisation).

5.3.2 Control Mechanism

The Edinburgh Prolog Blackboard shell (described in Section 2.7.2) was used for the implementation so that the control mechanism was already built into the shell. It consists mainly of an agenda, dynamically built during the consultation process. The agenda sequences the firing of the rules inside a knowledge module.

As already mentioned in Section 2.7.2.3, 'est' in the rules indicates the 'usefulness' of each rule and, thus, helps in building up the agenda. So, by giving appropriate 'est' values to the different rules, we may sequence the firing of these rules. The rule with the lowest 'est' value will be fired first. The rule with the next higher 'est' value after that and so on. This is the default methodology for conflict-resolution provided by the shell by giving numerical values to 'est'. However, the conflict-resolution strategy adopted in INDEX bypasses this approach and a new strategy was defined that suits the requirements of INDEX. As mentioned earlier in Section 4.5.4, the sequence of execution of the different knowledge-modules is as follows :

(ALTSEL -> STRANEX -> DETDEX -> OPTEX -> EVALUATOR)

Each of these modules consists of different sub-modules as discussed in Section 4.5.3. The sequence of execution of the sub-modules of ALTSEL is as follows :

(SYNTHESIS -> PREANA -> PREDES -> ECONOMICS -> DESIGN -> PREVALUATOR)

To accomplish this, the default method was to give simple numerical values to 'est' starting from the first rule of SYNTHESIS and increment them up to the last rule of PREVALUATOR. This, obviously, was not an elegant way of approaching the problem for the simple reason that if a rule was added to any of the modules at a later stage, all the 'est' values have to be changed for all the rules following it. Another reason was that the whole idea of modularity would get lost by this approach and the set of rules, in effect, would become one module instead of being broken down into sub-modules.

The second approach was to give symbolic 'est' values to the rules of the different

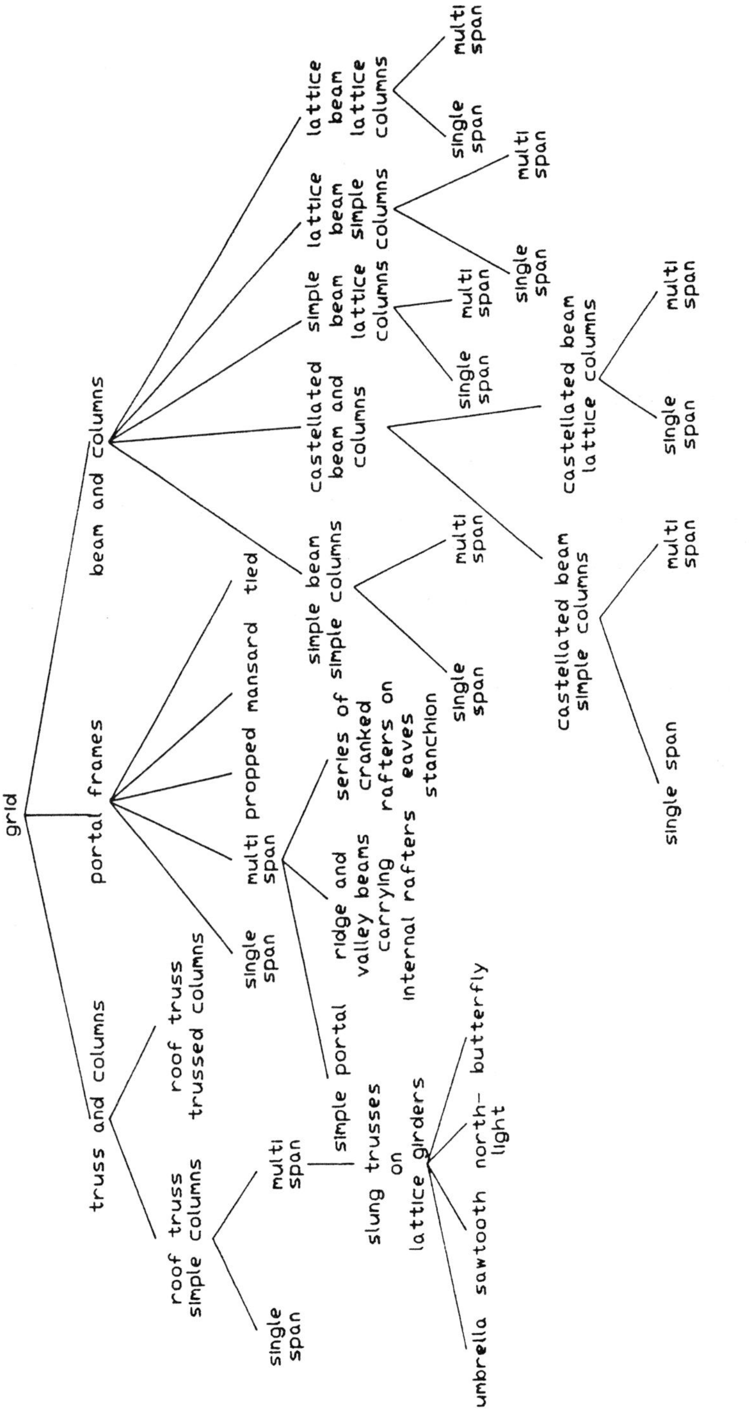

Figure 5.9 Solution search space for feasible lateral load resisting systems

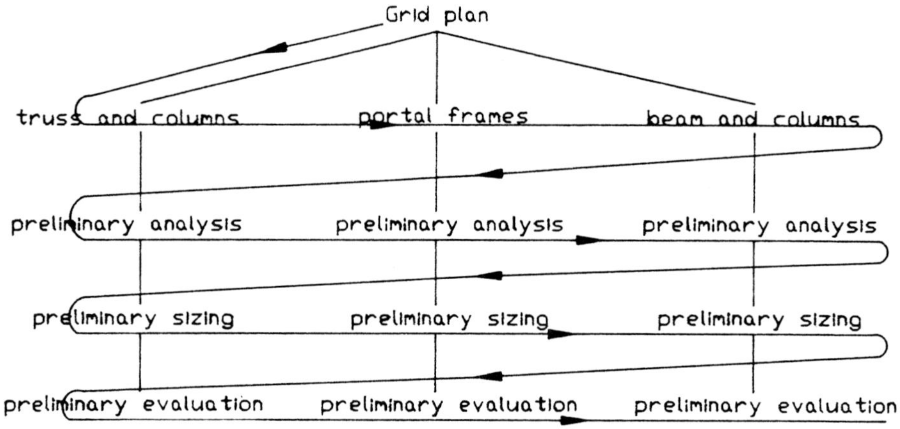

Figure 5.10 Problem-solving sequence adopted by ALTSEL

sub-modules specific to the rules of that sub-module only and define a different conflict-resolution strategy altogether. In this approach, the sequence of execution of the sub-modules had to be first defined and then the sequence of firing the rules inside each sub-module. A listing of a portion of this control mechanism is given in Appendix IV. The rules quoted earlier in this chapter have symbolic 'est' based on this approach. This approach avoids the problems of the default approach described above. To recap, following is an example rule from the SYNTHESIS sub-module using this approach:

```
if     [problem,span(X),true]
and     holds(X =< 60)
then    output_message('Single span portal frame is a feasible alternative')
to      add[synthesis,lateral_load_sys(single_span_portal),true]
est     synthesis(5).
```

The 'est' value of this rule indicates that this rule is the fifth rule inside the SYNTHESIS sub-module. The numerical value in this 'est' decides the firing of the rule inside that sub-module and the invoking of the sub-modules is decided by the top level conflict-resolution, as shown in Appendix IV. It is worthwhile to point out that this operation of sequencing of the firing of the rules as well as the knowledge modules according to the demands of the domain was made a lot simpler by using the Edinburgh Prolog Blackboard Shell.

5.4 Summary

The concepts involved in the development of ALTSEL, the preliminary design module of INDEX, were outlined. The *knowledge elicitation* as well as *knowledge representation* aspects of the development of ALTSEL's knowledge base were described in detail. Some implementation issues were highlighted with examples of some representative production rules. A description of the *control mechanism* suitable for the domain of this project was also given underlining the ease of accomplishing this using an expert system shell.

5.5 Conclusions

The following conclusions were drawn from the development of ALTSEL:

1. Artificial Intelligence tools and techniques provide a way of incorporating rules of thumb and heuristics into computer-aided design of structures with minimum effort.
2. *Protocol Analysis* was found to be a useful knowledge elicitation technique in the domain of design.
3. The *blackboard architecture* provided a flexible environment for the *propagation* of constraints between the different knowledge modules so vital for design.
4. The development of a fully-working system requires many different types of knowledge. Mere heuristics are not sufficient to solve real-life problems in structural design. The system needs to have numerical as well as logical capabilities.
5. Since a considerable number of decisions in the preliminary design stage are taken using heuristics, a system similar to the one described in this paper might perform satisfactorily in that domain. But, for the domain of detailed design, the system needs to have more capabilities, e.g., logical and mathematical inferencing from fundamental laws of structural engineering,

general knowledge of arithmatic etc.

References

1. **Welbank, M.A.**, *A Review of Knowledge Acquisition Techniques for Expert Systems.* British Telecom Research, Martlesham Heath, 1983.

2. **Gammack, J.G. and Young, R.M.**, *Psychology Techniques for Eliciting Expert Knowledge.* In Research and Development in Expert Systems, Bramer, M.A. (Ed.). Cambridge University Press, 1985.

3. **Bloomfield, B.P.**, *Capturing Expertise by Rule Induction.* The Knowledge Engineering Review, Vol 1, No. 4, 1986.

4. **Regan, J.E.**, *A Technique for Eliciting Categorical Knowledge for an Expert System.* Paper submitted to AAAI-87, 1987.

5. **Hayes-Roth, F., Waterman, D.A. and Lenat, D.B. (Eds.)**, *Building Expert Systems,* Addison Wesley, 1983.

6. **Mittal, S. and Dym, C.L.**, *Knowledge acquisition from Multiple Experts* , AI Magazine, pp 32-36, Summer 1985.

7. **Stefik, M., Foster, G., Bobrow, D.G., Kahn, K.M., Lanning, S. and Suchman, L.A.**, *Beyond the Chalkboard: Using Computers to Support Collaboration and Problem Solving in Meetings.* Paper submitted to CACM, 1986.

8. **Sriram,D.**, *Knowledge-Based Approaches to Structural Design.* Ph.D. Dissertation, Department of Civil Engineering, Carnegie-Mellon University, U.S.A., 1986.

9. **Horridge, J.F. and Morris, L.J.**, *Comparative Costs of Single storey Steel Framed Structures.* The Structural Engineer, Vol. 64A, No. 7, pp. 177-181, 1986.

10. **Maher, M.L.**, *Problem-Solving Using Expert System Techniques,* Expert Systems in Civil Engineering, Eds., Kostem,C.L. and Maher, M.L., ASCE, New York, pp. 7-17, 1986.

11. **Stefik, M.**, *Planning with Constraints - MOLGEN Part 1.* Artificial Intelligence, No. 16, pp. 111-140, 1981.

12. **Bates, W.**, *Introduction to the Design of Industrial Buildings.* Constrado, Croydon, 1978.

Chapter 6

DETDEX: The Detailed Design Module of INDEX

6.1 Introduction

This chapter describes the detailed design module of INDEX which is called DETDEX. As is well known, the bulk of the detailed design involves checking a design for conformity with various codes of practice. Accordingly, most of this chapter will include the description of different aspects of development of a generic standards processing sub-module of DETDEX which is called STAPRO. Checking a design against the relevant codes of practice is an essential part of structural design. The term codes of practice and standards are used interchangeably throughout the profession and refer to the same documents which prescribe a set of regulatory, mandatory or obligatory rules to which any design must comply. There have been attempts in the past to automate this process of checking a design against a standard. Before the advent of knowledge based systems technology, the approach adopted in most of this work was to hard-core the provisions of the relevant standard into the application program in the same language as the program itself. There are some inherent limitations of this approach and these will be discussed in Section 6.3. The emergence of knowledge-based systems technology has prompted an alternative approach to the process and some recent efforts (1,2,3) have produced good results using this approach.

As already mentioned in Section 4.5.3.2 the tasks before DETDEX are as follows:

1. to size the different components of the structure;
2. to select the applicable provisions of the codes of practice; and
3. to check the constraints prescribed by the codes as well as other soft constraints;

Thus, the bulk of the tasks performed by this module consist of checking a design based on the analysis carried out earlier as suggested by STRANEX to conform to relevant codes of practice as well as other constraints. The standards processing sub-module of DETDEX, STAPRO has been developed as a generic standards processor. The British standards for structural steelwork, BS5950, has been used to illustrate the use of the representation scheme developed for standards and also the generic nature of STAPRO.

6.2 Research in Knowledge-Based Detailed Design of Structures

Some of the important research works in knowledge-based approach to detailed design of structures have been already given in Chapter 3. Quite clearly, the bulk of the work in this area has concentrated upon knowledge-based standards processing. In the following sections, a detailed study will be presented which points out their limitations and drawbacks.

All the different research works in standards processing have been concerned with one or more of the following three issues:

1. standards representation;

2. standards analysis; and
3. standards manipulation

A brief discussion of each of these issues follows.

6.2.1 Standards Representation

More than a decade of research has identified the following four basic elements for representing any standard:

1. data items;
2. decision tables;
3. information networks; and
4. organisational system.

Each of these elements will now be described briefly.

6.2.1.1 Data items

Any variable found in a standard is called a data item or a datum. A data item may be of any of the following types:

1. boolean - can evaluate to 'true' or 'false', 'satisfied' or 'unsatisfied', etc;
2. numeric - can evaluate to a number;
3. multivalued - can evaluate to more than one values; and
4. single-valued - can only evaluate to a unique value.

Any standard can be thought of to be a collection of rules that specify the relationships between the data items. Each provision in a standard requires the evaluation of one or more data items.

6.2.1.2 Decision Tables

A decision table is a tabular representation of the rules in a standard. It consists of one or more conditions which can evaluate to either 'true' or 'false'. Based on the evaluation of the conditions, different actions are suggested. A decision table represents several rules, each rule suggesting a particular action for a particular combination of 'true' and 'false' values for the conditions. As is clear from Figure 6.1, a decision table has four parts. The upper left portion is the set of conditions, the upper right portion consists of the values ('true' or 'false') for each condition, the lower left portion consists of the set of actions and the lower right portion indicates the particular action to be taken for a particular combination of values for the conditions.

6.2.1.3 Information Network

An information network can be thought of as a tree whose nodes are the data items in a standard connected by arcs representing the precedence relationship between them. A data item is connected to other data items which are either ingredients or dependents. A data item having at least one ingredient is called a derived data item. In any network, there is at least one data item called the terminal data item which has no dependents. A data item with no ingredients is called a basic data item. Figure 6.2 is an example of an information network.

	Rule 1	Rule 2	Rule 3
Condition 1	True	True	True
Condition 2	True	False	True
Condition 3	True	True	False
Action 1	X		
Action 2		X	
Action 3			X

Figure 6.1 A decision table

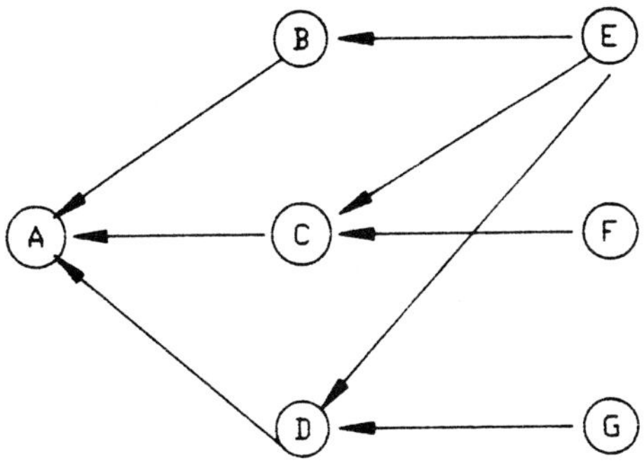

Figure 6.2 An information network

6.2.1.4 Organisation System

The organisation system helps in identifying the applicable provisions of a standard in any particular case. The provisions of a standard usually require the evaluation of certain constraints a design is supposed to satisfy. Any provision indicates the subject it concerns and the conditions to be satisfied. Thus, any requirement of a provision can be represented as follows:

<subject> must satisfy <constraints>

To facilitate the accessibility of any provision, all subjects covered by a standard are decomposed into different trees known as classifier trees. Classifiers are the basic element in the organisation system of a standard. A minimal classification system must have at least two classifier trees one for the physical objects covered and another for the conditions to be satisfied by the objects (4). Figure 6.3 is an example of a classifier tree taken from (4).

6.2.2 Standards analysis

The main issues concerning the analysis of standards are as follows:

1. consistency;
2. redundancy; and
3. completeness.

These three criteria are the main concern of the experts that compile a standard. The purpose of analysing a standard is to ascertain that it is consistent, it does not contain redundant information and the information contained in it is complete. However, to our knowledge, there is probably no system developed which analyses a standard in these lights.

6.2.3 Standards manipulation

Easy and effective standards manipulation is the ultimate objective of the earlier two areas of research in standards processing. This area concerns the actual use of a standard. Any standard in its existing form suffers from the following drawbacks which make its use as difficult as they are:

1. most provisions have ambiguous cross-references to other provisions, e.g. the provision numbers are not stated exactly and just a section number is provided;
2. there is no way to ascertain whether all the applicable provisions have been satisfied;
3. certain things are left to the engineer to decide;
4. the conditions and the actions themselves are sometimes not expressed clearly.

For proper use of a standard one of the primary objectives is to be able to pinpoint all the applicable provisions readily. Using the representational model described in Section 6.2.1, this task is performed by the classifier trees. As the rules of all the provisions are contained in the decision tables, the classifier trees help in locating all the applicable decision tables for any particular problem. Subsequently, the main task is to evaluate the conditions of the decision tables and suggest or evaluate the corresponding actions. This process has been accomplished in the past using various approaches. A good review of these approaches has been given in

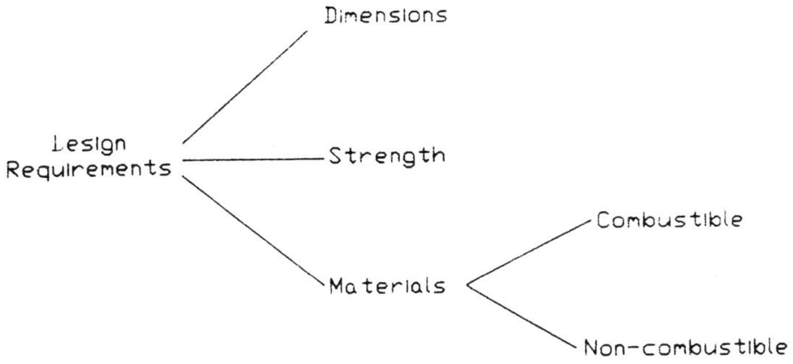

Figure 6.3 An example of a classifier tree (taken from (6))

(1,6). However, brief discussions follow:

Initially, the most popular approach was to encode the applicable provisions of a standard into the CAD program itself usually written in FORTRAN. This approach had many limitations (as discussed in Section 6.3). This approach was improved substantially by letting a preprocessor do the encoding of the provisions rather than doing it manually (6). Another program generates the decision trees from decision tables and produces source code in the form if IF-THEN-ELSE templates and then inserts the conditions and actions in the source code. This produces executable subroutines for subsequent evaluations of the decision tables. In all these approaches, the only difference lies in the level at which the evaluation of decision tables is initiated with a gradually increasing level of automation.

With the emergence of knowledge-based system technology, an alternative representation was tried in SPECON (5). The provisions of a standard were represented as production rules instead of decision tables. SPECON performed essentially the same functions as the earlier systems using a different representation formalism. Another difference was its dynamic rule-firing strategy instead of a fixed in the other cases. This approach of casting a standard as production rules also had its drawbacks (as discussed in Section 6.3).

6.3 Limitations and Drawbacks of Some Earlier Approaches to Standards Processing

Some of the important limitations and drawbacks of two of the popular approaches are being briefly discussed here.

6.3.1 Hard-coding approach

The biggest limitation of the hard-coding approach is the lack of flexibility as regards incorporation of any update to the standard in the program. The whole program will need to be drastically modified in order to incorporate any changes to the standard or even adding another provision from a standard. Another related limitation is the difficulty in including another standard not already encoded in the CAD program. Every limitation of this approach revolves around the lack of generic nature of the standards processor. In fact, in this approach there is no separate standards processor and the processing of standards is an integral part of the CAD program itself.

6.3.2 Production-rule approach

Most of the limitations of the hard-coding approach are overcome by representing standards as production rules. The whole program becomes very modular and, thus, adding or deleting a provision can be done without difficulty. However, this approach also suffers from a serious drawback in that it requires a very large memory space. Although casting standards as production rules is the most natural and intuitive way of representing a standard, it often results in a number of superfluous rules and one provision may have to be represented as more than one production rule. Even while casting the decision table entries as production rules, one is likely to end up with a large number of rules needing more memory than the equivalent decision table representation.

6.4 Nature of Standards Processing in Structural Design

Any standard meant to be complied with by any structural design is, by no means, complete for any particular design. By completeness is meant the complete set of requirements to be met by any design. There may be an equal (or even more) number of requirements not prescribed by a standard to be met by a design. It can, however, be argued that the standards are not cast with the objective of being a complete design guide. This point highlights one of the main differences between standards for structural design and some other codes such as building regulations. A building regulation code is complete in the sense that it prescribes the complete set of regulatory requirements to be met by a building. The other difference lies in the numerical nature of standards for structural design as opposed to the non-numerical nature of the building regulations codes. For these reasons, a system for processing a standard for structural design has to be fundamentally different from those for processing building codes (2) with additional capabilities. A system for processing standards for design will have to possess the following additional capabilities:

1. it has to possess knowledge other than that found in the standards; and
2. it has to have numerical processing capabilities.

To illustrate this point, let us consider Section 5.5.3.5 from BS5950 (7) that makes constant references to the tension and compression flanges of the rafter. The standard does not, rather, cannot specify the flange that will be in tension or compression in a particular case. To determine the state (tension or compression) of a flange, either a detailed analysis has to be carried out or, for approximate solutions, educated guesses can be made using heuristics.

6.5 Standards Processing for Structural Design

6.5.1 Classification of Clauses

For the purposes of organising the information contained in a standard, one of the primary tasks was considered to be the classification of the clauses found in a standard. For the purposes of this study, all the clauses are classified in the following three categories:

1. definition clauses;
2. application clauses; and
3. performance clauses.

Examples of these categories follow.

Definition clause

Section 5.1.2.3 of BS5950 states that "the notional horizontal load may be taken as 0.5% of the factored dead plus vertical imposed load applied horizontally". This clause defines the notional horizontal load and, thus, is classified as a definition clause.

Application clause

Sections 5.3.1 through to 5.3.7 of BS5950 specify the conditions to be met in order to undertake a plastic design. These sections specify the requirements to be met for the application of plastic design methods and, thus, are classified as application clauses.

Performance clause

Section 5.5.3.2 of BS5950 states that "horizontal deflection, d, calculated by linear elastic analysis at the top of any column due to the notional horizontal loading given in 5.1.2.3 applied in the same direction at the top of each column should not exceed h/1000, where h is the height of the column. In calculating d allowance may be made for the restraining effect of cladding." This clause prescribes a certain performance level of the structure as regards deflection and, thus, is classified as a performance clause.

A more comprehensive classification of clauses was proposed by Harris (4). He classified them into six categories, viz. basic, multiple, cumulative, application, synthetic and mixed.

6.5.2 Parts of a Clause

The purpose of any clause is to lay down certain criteria based on the satisfaction of certain other criteria. Thus, any clause consists of one or both of the following two types of criteria:

1. applicability criteria; and
2. performance criteria.

The applicability criteria of a clause are the conditions to be met in order that the performance criteria can be evaluated. An example of such criteria is Section 5.5.3.5 of BS5950 that states that "provided that:

1. the rafter is a UB section;
2. the haunch flange is not smaller than the rafter flange;
3. the depth of haunch is not greater than 3 times the depth of rafter;
4. the buckling resistance is satisfactory when checked as given in 4.3 using an effective length Le equal to the spacing of the tension flange restraints;

Lt may be conservatively taken as:

k1*ry*x/sqrt(72*x^2-10000) for grade 43 steel"

The conditions numbered 1 to 4 are the applicability criteria because they prescribe the conditions to be met before the formula given for Lt can be applicable.

The performance criteria are the prescription of certain performance levels to be achieved by the structure as regards various parameters if the applicability criteria are satisfied. The formula for the determination of Lt in the above is an example of a performance criterion.

The applicability and performance criteria of a clause are analogous to the applicability and performance conditions of the requirement decision tables of the representational model of standards proposed by Garrett (6). However, the difference is that in the model presented here, there is no reference to decision tables at all. The clauses are straightaway cast into facts without going through the intermediate stage of first casting the standards as decision tables.

6.5.3 Representation of Standard Clauses

As discussed in Section 6.2.3, the most popular representation of standards, hitherto, has been decision tables. Production rules have also been tried in SPECON (5). Another representation scheme is presented here which proposes to represent standards as *facts*. The clauses of a standard will be represented as **facts** rather than **rules** which is the most natural way of representing any standard which is effectively a collection of rules. Casting the standards as production rules was also experimented with before arriving at this representation scheme because of the following drawbacks observed in the production rule approach (also mentioned, generally, in Section 6.3):

1. rules were found to require much more computer memory than facts;
2. all the information in a standard need not be in rule form (e.g. definitions, tables), in which case the rules were found to be naturally inappropriate representation; and
3. generic standard-independent nature of the processor was found to be difficult to accomplish because of all the information being hard-coded in the rules whereas facts could be treated as data on which generic rules could operate.

It must be pointed out at this stage that although representing standards as facts rather than rules was developed independently, it is quite similar to the one proposed by Rasdorf and Wang (3). However, there are significant differences between the structure of the facts used in the two models. The most important ones being:

1. the nature and content of the information contained in the facts are different;
2. the facts in this model are created straightaway from the standards unlike Rasdorf and Wang's model which first casts the standards as decision tables and then converts the decision table entries as facts.

6.5.3.1 Types of Facts

Facts are either related to a particular clause or provision of a standard of general information about the standard itself. Thus, facts are divided into the following categories:

1. provisional;
2. organisational; and
3. general.

This categorisation is done at the highest level of abstraction and these will be further sub-divided at lower levels of abstraction in the next section. At this level, same terminologies have been used as Rasdorf and Wang (3) for the first two types for the sake of consistency.

Provisional facts

Provisional facts contain information relating to the criteria laid down by a particular provision of a standard. Based on the parts of a provision discussed in Section 6.5.2, the provisional facts are further sub-divided into the following two types of facts:

1. applicability provisional facts; and
2. performance provisional facts.

These two types of facts refer to the applicability and performance criteria respectively of the provisions, as described in Section 6.5.2 respectively.

Organisational facts

Organisational facts contain information regarding the precedence relationship between the different provisions of a standard. These contain information about each provision found in the standard.

General facts

These facts contain general information regarding the data-items contained in the code. These facts are used in generating explanations and are also needed in processing the standard, e.g., in finding out if a data-item is basic or derived, finding out the identity of a variable or the range of values that a variable could take.

6.6. Functional Details of DETDEX

6.6.1 Introduction

This section briefly discusses the functional details of DETDEX. Again, the focus will be on the implementation of a generic knowledge-based standards processor STAPRO, which utilises the representation scheme proposed in earlier sections.

6.6.2 Knowledge-Base of DETDEX

The knowledge base of DETDEX consists of the following knowledge modules (KMs):

1. CONS - this KM synthesises the constraints formulated by the DESIGN sub-module of the ALTSEL module described earlier in Chapter 5.
2. STAND - this KM contains all the standard-dependent knowledge i.e. all the provisional, organisational and general facts. Examples from this module are given in the following section.
3. STABILITY - this KM contains knowledge on the stability considerations for the design of portal frames. Following is a rule from this module:

```
if      notnow[(rafter,stability,req),_,true]
and     [eaves_haunch,present,true]
then    true
to      add[(rafter,stability,req),
            (
            (position,eaves_haunch(rafter_end)),
            (moment(0.87*mp))
            ),
            true]
est     rafter_stability(5).
```

   This rule specifies one of the general stability requirements for rafters and states that the value of moment at the rafter end of eaves haunch should be 0.87 times the plastic moment.

4. MEMPRO - this knowledge module contains knowledge about the general properties of the structural members, e.g. section area of different types of sections, section modulii, moments of inertia etc. Following is a rule from this module:

```
if      [section_type,rolled_ub_section,true]
or      [section_type,rolled_uc_section,true]
```

```
and   [section_dimension,web_thickness(Tw),true]
and   [section_dimension,web_depth(Dw),true]
and   [section_dimension,flange_thickness(Tf),true]
and   [section_dimension,flange_width(Wf),true]
and   [section_geometry,flange_area(Af),true]
then  Iy is ((2*(Tf/12)*(Wf^3))+((Dw/12)*(Tw^2)))
to    add[section_geometry,weak_axis_moment_of_inertia(Iy),true]
est   section_geometry(6).
```

This rule prescribes the formula to find out the area of a section if all its dimensions are known.

5. INTERPRETER - this KM contains knowledge on proper interpretations of different provisions in case of any ambiguities. This sub-module also contains heuristics to make guesses about some engineering criteria, e.g., if the loading type and their positions are known there are rules to infer the states of the flanges (i.e., tension or compression).
6. PROSOL - this module contains the set of rules that is responsible for processing the rest of the knowledge contained in the system. The rules in this KM are standard-independent. Such rules are sometimes called *control rules.*

Figure 6.4 shows the different modules of DETDEX. Out of the KMs described above, four sub-modules, viz., STAND, MEMPRO, INTERPRETER and PROSOL, constitute the standards processing part of DETDEX. In fact, these three sub-modules can be isolated to form a prototype stand-alone generic standards processor, STAPRO. The following sections describe the knowledge representation and other aspects of STAPRO.

6.6.3 Implementation of STAPRO

6.6.3.1 Knowledge Representation

The provisions of a standard are represented as facts instead of rules in the DETDEX. A provision is represented in the knowledge base as the following fact or entry on the blackboard:

```
[Criterion_type(Clause_number),(Data_item,[Ingredients],Expression,Operator),true]
```

An example is:

```
[perf('5.3.5'),(lm,[ry,fc,py,x],(900/((fc/130)+((py/275)^2)*((x/36)^2))^0.5),'=<'),true]
```

The fact above represents the performance criterion of provision 5.3.5 concerns the determination of the data item 'lm' whose ingredients are ry,fc,py and x, the actual expression for the evaluation of 'lm' is also given and the last argument of the facts is the operator '=<' which means that 'lm' should be less than or equal to the value of the expression.

Following is an example of the implementation of the applicability criteria of a provision. An applicability criterion can have two forms, viz,

```
[applicability(Clause_number),(X,boolean),true]
or
[applicability(Clause_number),(Data_item,[Ingredients],Expression,Operator),true]
```

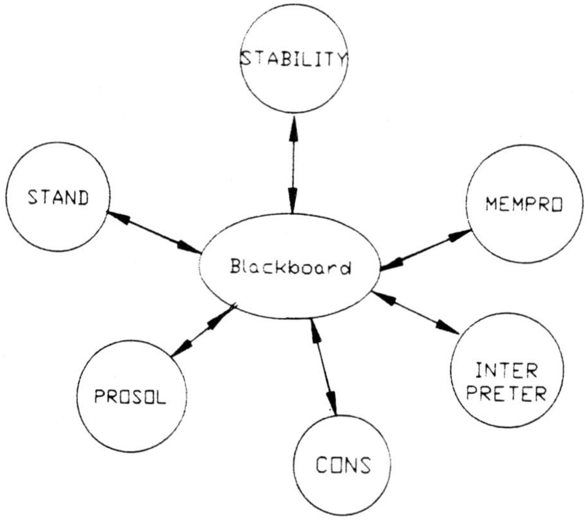

Figure 6.4 Knowledge-base of DETDEX

The latter of the two types of facts for storing applicability criterion of a provision concerns the evaluation of an expression to determine the applicability of a provision. The earlier one concerns the evaluation of an expression which can only take values either 'true' or 'false', e.g. rafter section UB, haunch_flange > rafter_flange etc.

Following is an example of an organisational fact:

[gen_des_req(Clause_number),(Object,Stresses,Limit_states),true]
or
[gen_des_appl(Clause_number),(Object,Construction_type,Design_type,Concerned_parameter),true]

The former fact represents that the provision number Clause_number concerns the design checks of Object under stresses Stresses and limit-states Limit_states. An object could be a structure or any part of a structure. An organisational fact for BS5950 is:

[gen_des_req('5.5.3.4),(portal_frame(column),_,(plastic,stability,whole)),true]

The '_' in place of Stresses implies that this clause applies for any stress acting on the object in question. '_' is called an anonymous variable in PROLOG.

The 'gen_des_appl' facts represent the general applicability clauses for a particular design and construction type. This means that these clauses have to be satisfied before the other clauses are checked. For example, for plastic design, loads have to be predominantly static. In case they are found to be otherwise (i.e., if any of these clauses is found to be violated), the process is stopped without going any further. Following is an example of such a fact:

[gen_des_appl('5.3.2'),(_,continuous,plastic,loading_type),true]

This fact represents the clause 5.3.2 of BS5950 which prescribes the loading type for the plastic design of any continuous construction.

In place of any of the different parts of these facts there could be atoms that may not exactly match any entry on the blackboard. Instead, there may be a name of a 'set'. For example, in place of Object there could be 'component' instead of a specific name of a component. This implies that the clause applies to any member of the set 'component' subject to the other parts of the fact being satisfied. To deal with this, there are other general facts stored in the knowledge-base, e.g.,

set_member(component,rafter).
set_member(component,column).

These facts imply that rafter and column are members of the set 'component'. Thus, the moment any member of the set in question is present, the condition for that particular clause to apply is met. This representation also allows the user to ask different questions to the system which will become clear from the run of the system given in Appendix V. Apart from the types of facts mentioned above, there are other facts that help the system in generating explanations (see Appendix V). Following are some such facts that help in identifying a variable and the possible range of values that it can take:

var_iden(k1,'constant depending on haunch depth/rafter depth').

```
data_item(k1,derived).
value_range(k1,'445 - 620').
```

6.6.3.2 Retrieval of applicable clauses

The retrieval of an applicable provision depends on the following five factors:

1. Object in question (i.e., structure, member or a secondary member);
2. limit states;
3. stresses acting on the object;
4. type of construction (i.e., simple or continuous); and
5. type of design (i.e., elastic or plastic).

All these items are input by the user. The questions asked, however, are not pre-programmed. Based on the type of object, design and construction type, the system searches through its knowledge-base to find the provisions that could possibly be applicable. It then asks for the limit states picking them up from the limit states part of the possible applicable provisions. The type of construction may be decided by the system itself in some cases, e.g., the system knows that a portal frame will be a continuous construction.

In case of structures, the system breaks the structure up into its constituent parts and then finds out the applicable clauses for each of them. Apart from that, it also finds out provisions applicable to the structure taken as a whole. Breaking up of the structure into its constituent parts is done on the basis of its knowledge about the structure composition. For example, the system knows that a portal frame is composed of rafters and columns in the simplest case. It will however, ask whether haunches are present as well because they may or may not be present.

6.6.3.3 Backtracking from a violated criterion

In case of violated criteria, the system can backtrack to suggest ways of correcting the violations (if asked to do so). The main concept being used here is building up dependency networks based on the ingredients of the involved data-item and their ingredients and so on until it reaches a leaf node which will be a basic data-item. The violation is then attempted to be corrected by manipulating one or more of these dependent data-items. The run given in Appendix V will illustrate the idea.

6.6.3.4 A brief description of a system run

The system runs in two different modes, viz

1. conformance; and
2. determination.

A solution tree is given in Figure 6.5. Based on the mode, the system takes different paths through the solution tree. In the conformance mode, there are three sub-modes, viz,

1. conformance of a data-item to the code of practice;
2. conformance to a particular provision; and
3. general conformance of a structure or a member of a structure to a code.

In the second case, the particular provision has to be supplied by the user. In the third

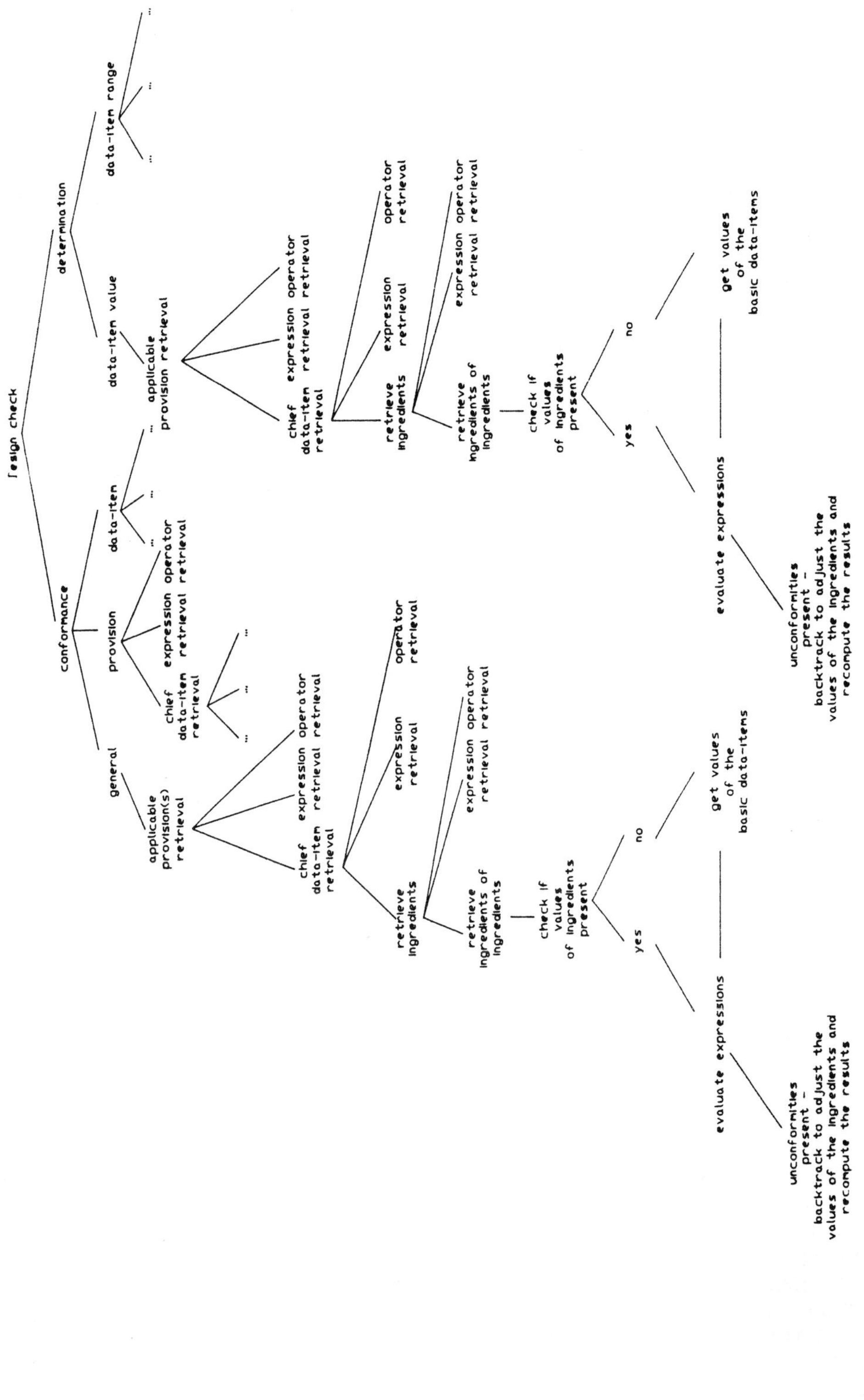

Figure 6.5 A typical solution tree generated by STAPRO

case, the system itself finds out the applicable clauses for the problem at hand using the general and organisational facts. Based on the involved data items and their ingredients a comprehensive network of all the involved data items is built and the checks performed accordingly. In the determination mode, the user has to supply the data item whose value or value-range has to be determined. In both the modes, the value of the ingredient data items have to be supplied by the user in case it is not already present on the blackboard. The system only asks the user to supply the values of the ingredients if it does not know about them, otherwise, it does not. In both the modes, there are additional explanation facilities (discussed in Section 6.7) provided to aid the user in case of any confusion or difficulty about anything being asked by the system. Appendix V gives some illustrative examples of different facilities and use of the system.

6.7. Possible Queries from DETDEX

In order to make the system more useful and powerful, additional useful querries have been provided in the system. A list of the useful ones follow:

1. show_gen_requirements - displays all the general design requirements;
2. show_applicable_clauses - displays all the applicable clauses of a standard;
3. advice(section(X)) - explains a particular clause, where X is the clause number;
4. explain(X) - explains any data item, where X can be any data item;
5. show_req(X) - displays the requirement at a particular location in the structure, where X can either be the co-ordinates of a point in the structure or any particular position (e.g. eaves);
6. show_location_of(X) - displays the positions of any particular secondary member or any part of a member in any specific state, where X can be any atom (e.g. rafter_tension_flange);
7. show_ingredients(X) - displays the ingredients of a data item if it is a derived data item, where X is the data item.
8. show_dependents(X) - displays the dependents of a data item X;
9. show_appl_cr(X) - displays the applicability criteria of a clause, where X is the clause number;
10. show_expression(X) - displays the expression to be evaluated for conforming to a particular clause X.

The use and utility of these commands will become evident from the illustrative examples given in Appendix V.

6.8. Summary and Conclusions

A description of DETDEX, the detailed design module of INDEX, was given. The bulk of the description dealt with the development of a generic standards processor, STAPRO. Representing standards as facts rather than decision tables or rules was proposed. Representing standard provisions as rules was found to be useful for many reasons. Facts offered a more concise and succinct way of representing standards. Representing standards as facts was found to be straightforward and did not need any knowledge of any programming language. Since the standards can be represented as facts, they can be very easily manipulated by a set of general standard-independent rules. In fact, programming in PROLOG only involves manipulation of facts by a set of rules (8). So, in that sense, the concept behind this type of representation and manipulation of standards is conceptually the same as programming in PROLOG. Furthermore, this representation allows for easier propagation of constraints between different parts of a system. As the *ingredients* of a *data-item* are represented explicitly in the knowledge-base, it is easy to infer the change in one data-item on the others. It would be difficult to enforce such a facility without such an explicit representation of the ingredients. This feature can be used to a great advantage in the whole design process (9). This point will be further explained in the next chapter. The set of generic rules responsible for manipulating the facts representing the standards were found to be general enough to be used with any standard represented using the representation scheme suggested.

References

1. **Fenves, S.J. and Garrett, Jr, J.H.** *Knowledge-based standards processing*, Artificial Intelligence in Engineering, Computational Mechanics Publications, Volume 1, Number 1, Southampton, pp 3-14, 1986.

2. **Rosenman, M.A. and Gero, J.H.** *Design codes as expert systems*, Computer-aided Design, Volume 17, Number 9, London, pp 399-409, 1985.

3. **Rasdorf, William J. and Wang T.E.**, *Generic Design Standards Processing in an Expert System Environment*, ASCE Journal of Computing in Civil Engineering, Volume 2, Number 1, pp 68-87, 1988.

4. **Harris, J.R. and Wright, R.N.**, *Organisation of Building Standards: Systematic Techniques for Scope and Arrangement*, Building Science Series NBS BSS 136, National Bureau of Standards, Washington, D.C., 1980.

5. **Sriram, D., Maher, M.L., Bielak, J. and Fenves, S.J.**, *Expert Systems for Civil Engineering - A Survey*, Report No R-82-137, Department of Civil Engineering, Carnegie-Mellon University, USA, July 1982.

6. **Garrett, Jr J.H. and Fenves, S.J.** *Knowledge-based Standards Processor for Structural Component Design*, Report No R-86-157, Department of Civil Engineering, Carnegie-Mellon University, USA, July 1986.

7. **BS5950 Part I**, *British Standards for structural use of steelwork in building - Code of Practice for Design in simple and continuous construction: hot rolled sections*, British Standards Institution, London, 1985.

8. **Clocksin, W.F. and Mellish, C.S.**, *Programming in Prolog*, Springer Verlag, Berlin, 1981.

9. **Chan, W.T. and Paulson, B.C.**, *Logic Programming to Manage Constraint-Based Design*, Microcomputer-Based Expert Systems in Civil Engineering, Ed., H. Adeli, ASCE Publications, New York, pp. 186-200, 1988.

Chapter 7

DESCON: The Design Reviewing Module of INDEX

7.1 Introduction

As discussed in Chapter 4, the main extension to the DESTINY model was proposed within the CRITIC module. The features lacking in CRITIC and the overall DESTINY model were also outlined in Chapter 4. To overcome the limitations of the DESTINY model, further extensions to its CRITIC module were suggested to take care of *non-monotonocity* often encountered in design. This chapter presents an approach to handling *non-monotonic* situations in structural design and describes the DESCON module of INDEX which incorporates the proposed solution to such situations. The main concept behind solving such *non-monotonic* situations is based on a network model of design as originally proposed by MacCallum (1). MacCallum's work centred mainly on numerical manipulation of relationships between design entities. DESCON uses a similar approach for symbolic, rather, than numerical reasoning.

7.2 Background

As already mentioned in Section 4.5.3, DESCON stands for DESign CONsultant. The tasks before DESCON are as follows:

1. to decide if the design or partial design is modifiable or a re-design is required when a change in specifications or violation of some constraints occurs;
2. to formulate the revised set of constraints in cases of a *modifiable* design;
3. to suggest the nature of modifications to be undertaken; and
4. to post constraints to the appropriate modules.

As is clear from above, the task of DESCON is to assist in solving problems that may arise during or after a design rather than doing the design itself. Problems arising during design could be innumerable and thus form a vast domain. A subset of such problems consists of situations arising due to modifications to parameter/s of a structure either deliberate or forced. It is this subset of problems that forms the domain of DESCON. A detailed description of such problems and relevant issues in the development of knowledge-based systems to tackle them may be found in (2). A shorter description is also given in the following sections.

7.2.1 Types and Possible Causes of Problems in Detailed Design of Structures

In its fundamental sense, detailed design of structures may be said to be the stage in the design process that involves the detailed analysis of the chosen structure from the preliminary design, the sizing and proportioning of its components and satisfying different constraints. A typical detailed design process may be seen to be the iteration of the following cycle :

'detailed analysis -> sizing and proportioning -> satisfying constraints'

Most of the detailed design stage consists of satisfying various constraints prescribed either in codes of practice or governed by other engineering or physical laws or other local constraints. It is the satisfying of these criteria that make the process so complicated since most of these criteria are prone to different interpretations. The structural components designed by the simple cycle given above, quite often, fail to provide a feasible solution in real-life situations and, thus, in such circumstances the detailed design process proves to be much more than merely going through that cycle. As already mentioned in Section 3.4, some common causes for such problematic situations arising could be :

(1) some 'local' constraints not considered,

(2) mis-interpretation of the provisions of design codes or

(3) ignoring certain provisions of design codes.

In many cases, the whole structure may either have to be re-designed or even a completely different alternative may have to be considered. In the following sections, some examples will be given from real-life situations to highlight the difference between a theoretical and a practical solution to a problem. Steps in the solution to some of these problems will be identified with the knowledge required at every stage. Examples will be taken from the area of plastic design of portal frames. The reason for selecting portal frames as an example is that although they are one of the simplest type of structures, there are many criteria to be satisfied.

7.2.1.1 Types of Constraints

The types of constraints that one normally encounters in structural design could be classified into two broad categories :

(1) Internal constraints and

(2) External constraints.

Internal Constraints - are constraints that arise due to the interaction between the constituent parts of a structure.

External Constraints - are the constraints imposed by objects or entities external to the structure.

These constraints can be further sub-divided into the following different categories (3,4) :

1. Designer constraints - these are imposed by the styles of an individual designer.
2. Client/User constraints - these are the constraints laid down by the client, which are basically tailored towards the end use of the building.
3. System-based constraints - these are structural system based constraints.
4. Regulatory constraints - these include the constraints imposed by the different regulatory governmental agencies such as the codes of practice.

The different constraints under any of the categories are equally responsible for making the design process as complex as it is. It is difficult to attribute the complexity to one category in particular. The examples quoted later on will show how the different constraints under the different categories make a designer's life difficult on different occasions.

From another point of view, the constraints may be divided into the following two broad categories :

1. Hard constraints - these constraints are the ones which are rigid and it is absolutely essential to satisfy them. The constraints imposed by the codes of practice fall under this category.
2. Soft constraints - these constraints are not as rigid as the hard ones and can be relaxed. For example, the designer may find that certain parts of a building require beams of greater depth than the constraint imposed by the client. In such cases, he can relax the constraint imposed by the client on the depth of the beam.

Although the constraints falling under any of the above-mentioned categories pose problems on different occasions, all of them have distinct features that make their incorporation into computer-aided design software increasingly difficult. Some such features of the soft constraints are :

(1) they are innumerable,
(2) they are difficult to anticipate,
(3) they are dynamic (i.e., they change with place, time, intended use of the structure etc.).

Some such features of the hard constraints are :

(i) they are prone to different interpretations,
(ii) they are dynamic, they are specialised in nature, i.e., they need specialist knowledge to tackle them.

It is these characteristics of both (soft and hard) the types of constraints that are mainly responsible for frustrating any attempts to encode them in computer-aided design software. From the point of view of incorporating these constraints in software, both pose different types of problems. As far as soft constraints are concerned, it is practically impossible to capture all the different constraints that one might encounter in real-life. In fact, the hard constraints pose less of a problem as far as their incorporation in computer programs is concerned. But, there again, the problem is that their solution (as we shall see) needs 'knowledge' that could be :

(a) vast,

(b) conflicting,

(c) uncertain,

(d) highly mathematical etc.

The emergence of knowledge-based approaches seems to offer some solution to the problems mentioned above. The only way to develop a knowledge-based system for the design of structures appears to be to attempt the capture of how experts approach problems in their day-to-day practice. After all, an expert does not have all the possible 'soft' constraints in his mind either and still he is able to solve problems arising due to them whereas we find it so difficult to enable the computers do it. One thing that immediately captures one's notice is that none of the computer programs have all and different types of 'knowledge' that these experts use to solve real-life problems.

Let us examine the detailed design of structures a bit more closely. As stated earlier, the bulk of the detailed design stage consists of checking of different design criteria. There is detailed guidance available on general principles of design (6). But, the engineers have to interpret this guidance properly. This requires a thorough understanding of the principles of design as well as their practical implications. It is a matter of common knowledge that most engineers either tend to misinterpret this design guidance or even ignore it completely, e.g., questions like *'what is the purpose of the eaves member in a portal frame ?'* are quite common which means that someone appears not to be using eaves members (6). For definition purposes, the detailed design stage may be broken down into the following three sub-stages :

(1) analysis,

(2) sizing and proportioning and

(3) checking.

A failure at sub-stage (3) could necessitate another 'analyse-size-check' cycle. However, in addition to this, a structure which has satisfied all the different criteria might not still be able to face the 'actual' conditions at site due to various reasons. This will necessitate a completely new 'analyse-size-check' cycle or the designer may even have to go back to the preliminary design stage and pick up a different alternative altogether.

To illustrate the above-mentioned fact, let us consider a simple but quite common example. One of the criteria in the design of portal frames is that the members have to be checked for stability, both lateral as well as torsional. To give lateral supports to the columns is almost a must in any case. Quite often, it might not be possible to restrain a column by sheeting rails or whatever over its full length. One common reason for this could be an opening (say, for a door) adjacent to the column. This is a simple but a typical 'actual' condition which might render a completely safe and properly designed and checked structure useless. These cases require ingenious solutions. The designer has to use his experience or approach the problem from first principles to solve such problems. Mere knowledge of 'analyse-size-check' cycle will be of little or no help in such cases. To approach a problem from first principles, one has to have a very thorough and clear understanding of the theoretical aspects of analysis and design. One of the simple solutions to this problem would be to increase the column size to a suitable section until it becomes stable over its full length. Although this might not even take a couple of minutes for the experienced designer to arrive at this solution, the amount of knowledge and experience contributing to this solution is not little. Figure 7.1 is a sketch of the different steps in reaching this solution. Alongside, the 'knowledge' used in moving from one step to the next is also identified.

7.2.1.2 Some illustrative examples

In the following paragraphs, some examples are given from real-life situations mainly concerning the stability criteria of the plastic design of portal frames. The main problem as far as stability considerations are concerned is to check whether a member is stable or not. In case of a member being unstable, the decision has to be made as to how many lateral supports are necessary and where they should be located in order to make the member stable. There could be many ways of approaching this

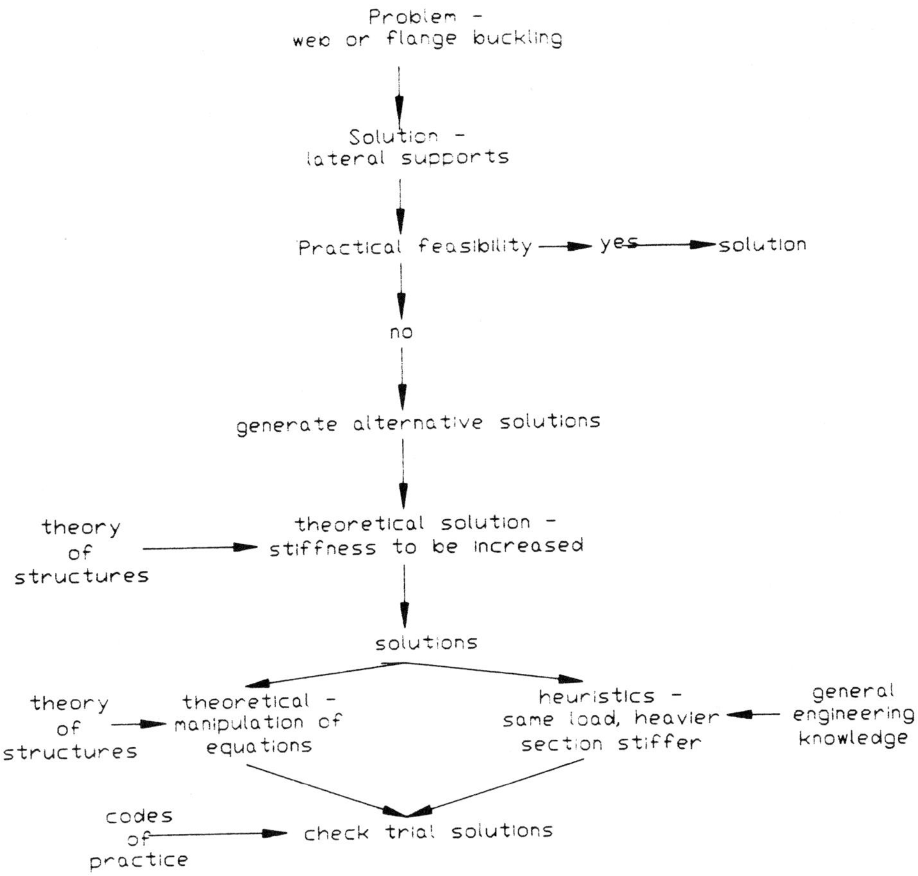

Figure 7.1 Steps in the solution of a simple design problem alongwith the contrbuting knowledge sources

problem. As far as checking whether a member is stable or not is concerned, it is a matter of checking the appropriate provisions of the code of practice. These will, usually, be checking some conditions given by some equations. Although this sounds simple enough, an inexperienced engineer might very easily run into rough waters. Here is an example: British standard BS5950 prescribes that the lateral restraints should not be placed at more than D/2 from the plastic hinge position, where D is the depth of the section. It does not say a word about the direction in which this distance (D/2) should be measured. This distance could be interpreted to be in the transverse direction to the flange across the cross-section or it could be taken to be measured along the flange. This example is taken from a real-life situation (7,8). This querry was asked by quite an experienced engineer in response to (6). The repercussions of not placing the restraints at the right position could be disastrous. This simple example clearly shows how the provisions of the codes of practice are vague and, thus, prone to mis-interpretations. The correct answer to the above-mentioned problem is that the distance D/2 should be measured along the flange. The reason is that it is the flange that is prone to lateral displacements. In fact, the ideal position of the restraint would be at the hinge itself. But, the above-mentioned rule is devised to take care of situations when it may not be possible to place the restraint exactly at the hinge position. The expression D/2 is based on results of experimental and theoretical research. D/2 on either side of the hinge is found to be the 'danger zone' that needs attention. A related problem could be what happens when it is not possible to put the restraints within D/2 on either side of the hinge. This is a similar situation to the one mentioned earlier (door beside a column). One simple solution to this problem could be to remove the chosen section and increase the section size until it becomes stable over its entire length (same solution as before). Figure 7.1 has already shown the steps and knowledge contributing to this solution. Another solution to this problem might be to encase the column upto the underside of the haunch. Still another solution could be to have brickwalls physically tied to the column (6). However, one problem with this solution is that unless the brickwalls are physically tied to the steel column by means of studs or wall ties, it is *doubtful* whether such walls could stabilise a column in the lateral direction over the life span of the building (6). This example illustrates another feature of the knowledge required to solve problems in structural design, i.e., that some knowledge is *doubtful* or uncertain.

Another example to illustrate the conflicting knowledge quite often encountered in structural design is with respect to lateral restraints. It is recommended that it should be the compression flange that needs to be restrained from lateral displacements and not the tension flange. The reason being that it is the compression flange that is highly stressed. However, in practice, many designers use tension restraint or restraints in pairs, i.e., both compression and tension restraints. Although this defies the recommendation, it works perfectly well in a number of cases. In fact, in some cases, it becomes a necessity. This example is also taken from (7). The question asked was,' On the question of lateral restraints and the observation that the strength requirements of 4% flange area and l/r of 100 produce similar results - this will apply only to strut stays. Many designers use tension restraints, tension stays or stays in pairs, i.e., tensile and compressive restraints.' The author's reply was,' I agree that the recommendations for the lateral restraints apply to stays sustaining compression load. The questioner's comments that seem to imply that, because stays are usually used in pairs, there is no need to apply such recommendations as the tension restraint will prevent movement. Under ideal conditions, that might be the case, but such conditions do not exist on site. Lack of fit, hole clearance, axial deformations all allow the compression flange to move sufficiently to induce local buckling or even lateral-torsional instability. Thus, the tension stay may not restrain the rafter adequately so as to prevent lateral movement, especially those stays that are positioned at an inclination to the horizontal more severe than 45 degrees. It should be borne in mind that the restraint has to be 'complete', i.e., not allow any lateral movement of the compression flange. If the cladding system has not sufficient out-of-plane rigidity, i.e., allows vertical movement (up or down) of the secondary members, then lateral movement of the compression flange may occur. It also depends on the moment resistance of the secondary member to main-member connection. Tests have shown that the magnitude of the compression force in the lateral restraints is of the order of 1.5 % of the squash load of the flange being restrained. However, the real criterion is stiffness, but *it is difficult to formulate a simple rule*. Neverthless, l/r < 100 is suggested. In addition to the restraining force, the lateral stays attached to the roofing system will attract a certain amount of load via the purlin, as it will, in effect, be acting as a prop to the purlin.' Here is another example. It is recommended that the haunches should not be designed to contain hinges. Another question from (7,8) is quoted,' It is recommended that the haunches are not designed to contain hinges. Is there any reason why the recommendations in (9) should not be used, provided that all the necessary lateral restraints are present, checks carried out as (9) and hinges effectively restrained ?' The author's reply was,' With the proportions of a typical haunch, a plastic hinge in the haunch (if allowed) would tend to cause yielding throughout the haunch. This would probably necessitate more restraints but also increase the possibility of web buckling in the haunch leading to premature failure. Therefore, I would still recommend that the haunch remain elastic. It should be noted that, even in the 'elastic' state, a certain amount of premature yielding in the haunch is present because of residual stresses.' These two examples clearly show that the designer has to, often, encounter knowledge that is *conflicting*. Another feature that also becomes clear is that the knowledge required is vast and includes the effects of interaction between different parts of structure. Still another feature that becomes clear by these examples is that, on many occasions *it is difficult to formulate simple rules* to solve problems.

Based on the foregoing discussions, it is easy to infer that the two basic types of knowledge required to solve problems in structural design are drawn from :

1. theory of structures and

2. design codes.

Both these types of knowledge will now be discussed in a bit more detail. The first thing that one needs to take into account is that both are often inter-dependent. Thus, both the types will be discussed together. Codes of practice are meant to be a set of rules based on the results of experimental and theoretical research. These rules are supposed to be followed as strictly as possible in actual design practice. However, the biggest problem with these rules is that their implementation in practice depends largely upon an individual designer's interpretations. Some examples were cited from real-life situations to highlight this point. These interpretations are largely dependant upon the individual designer's knowledge and understanding of the principles of analysis and design of structures as well as his experience. The knowledge based on his experience basically consists of heuristics acquired over the years.

As far as the knowledge on the theory of structures is concerned, it could be divided into two categories :

1. Theoretical knowledge consisting of different theorems of the analysis of structures. This could include the knowledge of material properties and behaviour of different types of structures under different types of loading

conditions.

2. Heuristics or rules of thumb acquired by experience. This will also include the 'general knowledge' about structures, e.g., what is a portal frame, what is the effect of varying cross-sections on the behaviour of structures etc.

The particular areas or occasions of application of one or the other type of knowledge mentioned above is practically impossible to anticipate. Perhaps, it would be right to say that whenever one's heuristics or "general knowledge" fail to explain a particular situation, one should take recourse to the theoretical knowledge. These problems might be the ones that need approaching from the first principles.

As is clear from Figure 7.1, some problems can be solved either by using 'general engineering knowledge' or theoretical knowledge. For example, for the problem tackled in Figure 7.1, the use of theoretical knowledge will mean manipulating equations and inferring the effect of varying one parameter on the others. This may not be an easy task to incorporate in a system. Although this problem was easily tackled by either of the approaches, it might not always be the case. Some problems will almost certainly require to be solved by the analytical or theoretical knowledge. On the other hand, some problems may be very easily solved by just using the 'general engineering knowledge'. Clearly, the problems that can be solved using simply heuristics are easier from the point of view of incorporating them in a knowledge-based design system. The other class of problems undoubtedly pose bigger problems for developing a knowledge-based system. Figure 7.2 shows a schematic model of the different knowledge modules that a knowledge-based expert system for detailed design should consist. It also shows the characteristics of knowledge inside each module.

The examples cited in the preceding sections were mostly taken from (7,8). The questioners included some very experienced designers of portal frames from all over the U.K. The author of the original paper (6) to which all the querries were directed is himself a very experienced and well-known researcher in the area of plastic design of portal frames. The discussions clearly show the amount of disagreement among even the very experienced designers. This, in turn, indicates that there is nothing *absolute* in the world of structural design or perhaps any design in general. The problem then is how to capture such conflicting and uncertain pieces of knowledge and infer sensibly from them.

7.3 Development of DESCON

7.3.1 Introduction

Based on the preceding discussions, a prototype capable of helping in the types of problems discussed earlier, was developed. The following paragraphs give a brief overview of the prototype. It is called DESCON (DESign CONsultant).

7.3.2 Problem-Solving by DESCON

7.3.2.1 Basic Strategy

As discussed in Section 7.2, one of the first things DESCON has to decide is whether or not the design or partial design is *modifiable* or *redesignable*. DESCON accomplishes this by first identifying the type of the involved parameters. A parameter or design entity could be either an *independent* (or *basic*) or *derived* as defined in Section 6.2.1.3., e.g., span is a *basic* entity whereas section area is *derived*. If a basic entity is involved, the remedial action suggested by DESCON is simply *re-design*. This is the simplest case. If a *derived* entity is involved, the decision has to be an

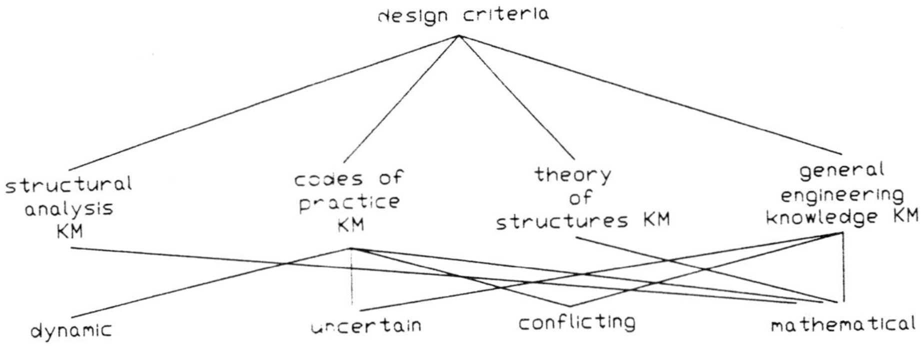

Figure 7.2 Schematic model of the knowledge modules for a knowledge-based system for the detailed design of structures alongwith the characteristics of knowledge within them

informed one. In such cases, the decision is guided, to a considerable extent, by the knowledge contained in DESCON. Under this category too, a relatively simpler case is when there are large differences in the *assumed* and *actual* properties of the structure. However, even in this case, the exact definition of large is an informed decision taken on the basis of some heuristics. In cases when a provision of a code is not satisfied (this could be the case of difference in *assumed* and *actual* properties), the standards-processing sub-module, STAPRO, also has the capability to suggest remedial actions based on dependencies between the involved data-items by backtracking within itself. As discussed in Section 6.6.3.3, STAPRO finds the dependencies by finding out the ingredients of a data-item which are explicitly stored in the knowledge-base. But, in other cases when the specifications of the design are changed, the solution process is much more complex. DESCON's real utility can only be evaluated in solving these cases. An example of such problems is the 'door beside a column' problem mentioned in Section 7.2. A description of how DESCON solves such problems is given in the following sections. The main concept behind DESCON's problem-solving strategy is based on the *network model of design* as proposed by MacCallum (1). MacCallum implemented a system called DESIGNER (1) based on this model.

7.3.2.2 Network Model of Design

A brief description of MacCallum's network model of design was given in Section 3.3.2. A more detailed discussion follows. The network model of design is based on the understanding of the relationships between different design entities. MacCallum suggests that in order to be able to explore alternative design concepts intelligently, a thorough understanding of the relationships between the different participating entities is of paramount importance. A designer must be able to predict the effects of change in one entity on the others. He used this idea to create new design concepts rather than improve or modify existing ones. However, he did recognise its utility in improving or modifying existing designs as well. The work presented here in the form of the knowledge-based prototype, DESCON, concerns using the network model to modify or improve existing or partial designs rather than create a new design altogether.

In the network model of design, the design process is represented as a *directed network*. In such a network, each node represents a design entity and an arc the dependency between two entities. The direction of the dependency is indicated by the direction of the arc shown by the arrowhead. Figure 7.3 is an example of a *directed network* representing the relationships between the entities A, B, C and D. The network represents a relationship defined (possibly) by a formula. The utility of any such network in inferring the effects of one entity on the others is not difficult to imagine. When traversing the network in an opposite direction than the one indicated by the arrowhead, the influence in an opposite sense can be inferred. For example, to modify D, it is easy to infer that A, B or C needs to be modified. On the other hand, if one of A, B, or C is changed, the corresponding change in D can also be inferred. Thus, apart from just finding out the value of D from A, B and C, the effects of change in both the directions can be inferred. However, real-life networks are much more complex than the one shown in Figure 7.3. In real-life situations, a network may involve many more entities and the relationships may not be just the direct ones but there may exist an equal or even more number of indirect relationships. In such cases, it becomes crucial to examine the indirect effects of change in one entity on the others as well. This may not be an easy task. Figure 7.4 is a network for the design of a simply supported beam. It is clear that although the cost is not directly related to the span, they do have an indirect relationship which may be important to consider. To get an idea of the number of dependent entities in structural design, Figure 7.5 is an abstraction hierarchy of the design entities commonly involved in a design problem.

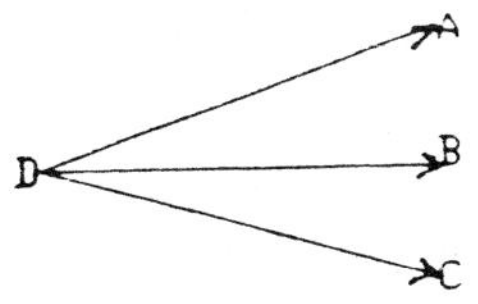

Figure 7.3 A Directed Network

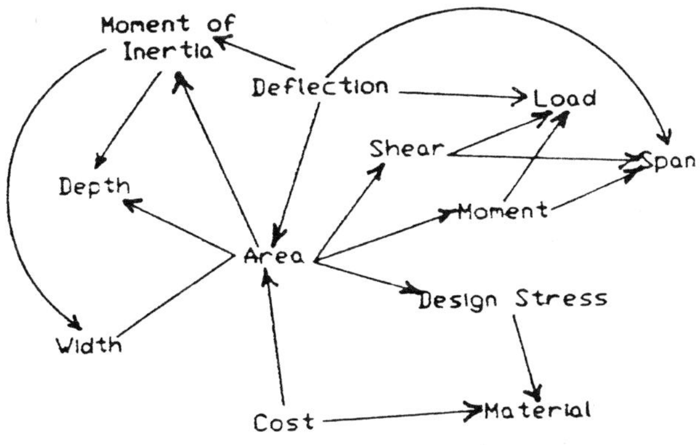

Figure 7.4 A Dependency Network for the design of a simple beam

Quite clearly, a system capable of handling all the entities shown in this figure must have a very large knowledge-base and DESCON's knowledge-base is not yet large and comprehensive enough to deal with such a multitude of entities.

7.3.3 Differences between DESIGNER and Other Similar Systems and DESCON

As mentioned earlier, DESCON is different from DESIGNER in that it is not supposed to create a new design concept. It is invoked whenever there is a change in specifications, either forced or deliberate, or a violation of constraints occur. The purpose and need for such a knowledge module in a design system has been explained in Section 4.6. The cases in which such a system may be required has also been explained in Section 7.2.1.1.

Another system based on MacCallum's network model of design is RELATOR (10). RELATOR uses the network model in much the same way as DESIGNER. It infers the effects of change in one design entity on the others by manipulating relationships between the dependent entities. These effects inferred by RELATOR are purely of numerical nature. For example, if the depth of a beam is changed, it infers the change in concrete volume, area of steel etc. In contrast, DESCON's task is centred more on *non-numeric, symbolic inferencing*. For example, consider the simplified network of Figure 7.6. DESCON's task is to infer the *conceptual* influence of one entity on the others in the network. For example, if the problem is to find alternative ways of stabilising a column because the chosen way is infeasible for some reason, it will infer that the stiffness needs to be increased and finally it concludes that the section area must be increased in order to do that. It is thought that once the *conceptual* decision is taken, the rest is a relatively simpler process and involves wholly or mostly numerical computations using well-defined formulae.

7.3.4 Artificial Intelligence Techniques used by DESCON

The principal Artificial Intelligence (AI) technique used by DESCON is dependency-directed backtracking (DDB). DDB (11) has already been discussed in Section 2.3.6.5. It involves keeping track of the *antecedents* of the *consequent* decisions taken by the system. This record is used to resolve any contradictions encountered later in the solution process. Whenever any such contradiction is encountered, the *antecedents* of the *consequent* decisions affected by the contradiction are deleted from the global database. This involves discarding all the inferences made based (directly or indirectly) on those *antecedents* to be deleted as well. In this method, only the involved *facts* are deleted whereas in the other form of backtracking (*chronological*), everything is deleted from the database indiscriminately. DDB also provides a way of generating useful explanations using the records of involved *antecedents* and *consequent* decisions.

Apart from DDB, DESCON also uses specific heuristics to find solutions in cases where DDB cannot find one or else to generate a number of alternative solutions directly without taking recourse to DDB as discussed in the next section.

7.3.4.1 Backtracking and Manipulation of Relationships in DESCON

In the context of DESCON or any blackboard system, the global database is the *blackboard*. DDB in DESCON is accomplished by keeping a record of the *Tag No.* of the *antecedent* and *consequent* entries. The *Tag Nos.* are explained in Section 2.7.2.3. This is done by asserting the following unit clauses in the internal database as the solution proceeds:

supports(Tag1, Tag)

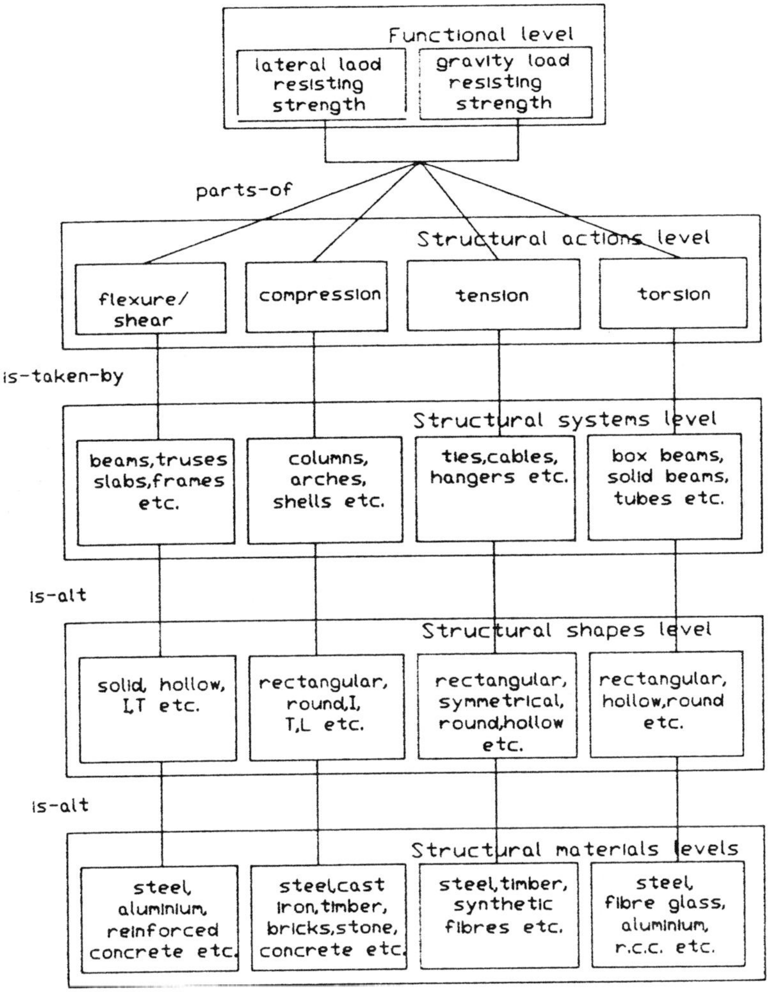

Figure 7.5 An abstraction hierarchy of structural design entities

Tag1 and *Tag* are the identification numbers of the *antecedent* and the *consequent* entries respectively. Based on this information, the deletion of any entry on the blackboard results in the deletion of all entries dependent on it. This plays an important role in the consistency maintenance of the *blackboard* as discussed in Section 2.7.2.4.

The information contained in the above-mentioned 'supports' clauses is used by DESCON to build the dependency network of the involved entities. Apart from finding the dependencies in this manner, DESCON's knowledge-base also contains explicit representation of the dependent entities in structural design (Section 7.3.5.2). In line with the network model of design, it uses these dependencies to manipulate the relationships between the entities in order to effect the desired changes in them. The relationships between the entities are explicitly stored in the knowledge-base. The relationships between the involved entities are manipulated by DESCON using certain general principles depending on the type of relationships discussed in Section 7.3.5.4. Backtracking is mainly used in DESCON to find out the *antecedents* of the *consequent* decisions invalidated by the change in specification or violation of constraints. For example, if it is required to discard one of the purlins, all the supports for having a purlin can be found out on backtracking. These supports become the ultimate goal to be achieved by DESCON, e.g. if a purlin is removed it has to find out alternative ways of stabilising the rafters and also supporting the roof. In this way, DESCON identifies the entity to be remedied and also the desired change in it. In the example cited here, the entity in question is the stability of rafters and the desired change in it is an *increase* and thus, the ultimate goal to be achieved becomes 'increase stability'.

An important decision DESCON has to take before manipulating relationships between entities is to identify the entity to be manipulated out of all the dependent entities that it comes across in the traversal of the chain of dependent entities (Figure 7.7a) stored explicitly in its knowledge-base. It does so by applying some heuristic tests to each dependent entity it comes across while traversing the chain of dependent entities. In DESCON, the test is to ascertain if the entity is one of the basic parameters of the object in question. For example, if the object in question is a member of the structure, any entity related to its section properties is considered a basic parameter, i.e., section dimensions, section area, section modulus and moment of inertia. Once the entity to be manipulated is identified, the next task before DESCON is to ascertain the nature of manipulation to be undertaken. It accomplishes this by manipulating the relationships between the original entity in question and its dependent entity found by either backtracking or traversing the tree of dependent entities. In some cases, there may not be a direct relationship between the original entity in question and the other entity selected by DESCON to be manipulated in order to remedy the situation. Considering Figure 7.7b, if entity A is the original entity in question and the entity to be manipulated to remedy the situation is D, there is no direct relationship between them. In such cases, DESCON first infers the change in B (which is directly dependent on A) that will effect the desired change in A. Subsequently, it infers the change in C that will effect the desired change in B and so on until it reaches D. Thus, in cases of indirect relationships between entities, DESCON has to manipulate relationships between the directly dependent entities down the chain of dependent entities until it reaches the final entity to be manipulated.

7.3.4.2 Heuristic Solutions

DESCON does not always have to find solutions by manipulating relationships between dependent entities. In many cases, it uses heuristics to arrive at a solution. DESCON's heuristics mainly consist of rules about the effects of some design entities on the the behaviour of a structure. DESCON infers the effects of the removal of

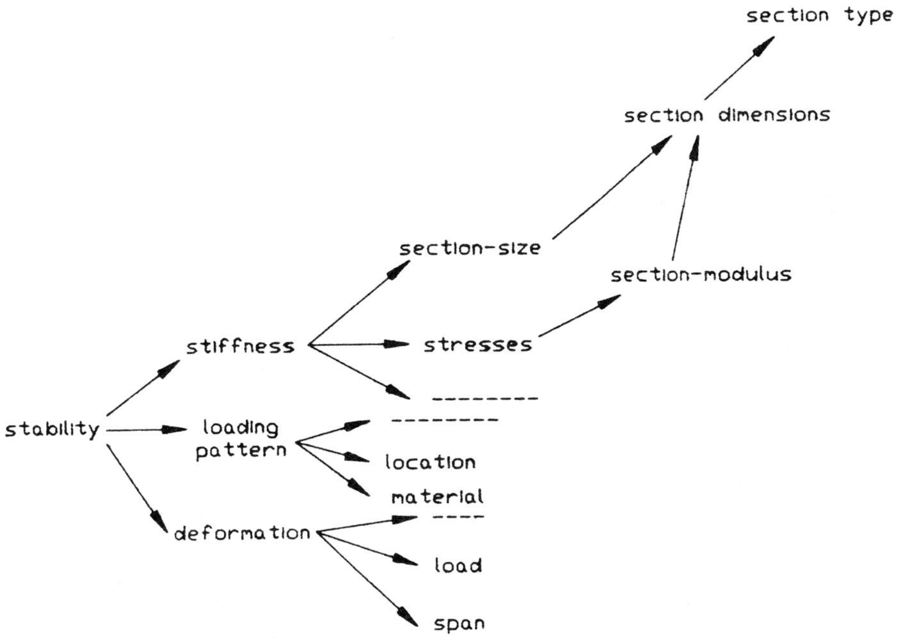

Figure 7.6 A portion of the dependency network

some design entity by finding the opposite of the effects of its presence. For example, if a secondary member is removed, it can immediately infer that the stability of the supported member is decreased if the secondary member's effect was to stabilise the member. Section 7.3.5 contains a description of the knowledge modules containing such information. This information is also used directly to suggest remedies to some situations, e.g., excessive deflection in a portal frame may be remedied by introducing a ridge haunch.

7.3.4.3 Types of Relationships handled by DESCON

DESCON can handle the following types of relationships:

1. directly_proportional -
   - constrained;
   - unconstrained.
2. inversely_proportional -
   - constrained;
   - unconstrained.

As their names suggest, the *directly_proportional* relationship implies that a change in one entity will effect a similar change in the other dependent entity. The *inversely_proportional* relationship implies that a change in one entity will effect an opposite change in the other. *Constrained* relationship signifies that the relationship between two entities involves some other entities as well. Thus, the effect of a change in one of the entities on the other can be only inferred if all the others are not changing. *Unconstrained* relationship signifies a direct relationship between two entities when no other entities are involved and the effect of change in one on the other can be inferred directly. At the moment, DESCON can only handle *unconstrained* relationships or only such *constrained* relationships in which only the two entities in question change and the rest remain constant. The problem of many entities changing simultaneously comes under the domain of *multi-variate analysis* and is beyond the scope of this project.

7.3.4.4 Passing the Control to the Appropriate Module

As mentioned in Section 7.2, DESCON's task also includes passing the control of execution to the appropriate module once it has decided about the nature of modifications to be undertaken to a design. This operation is accomplished utilising the features of blackboard architecture. In most cases, the blackboard architecture helps in accomplishing this with minimum effort. The *entry* that is found infeasible due to the change in specifications is simply deleted from the blackboard. As discussed in Sections 2.7.2.4 and 7.3.4.4, this leads to the deletion of all the *entries* that depend on this entry. This is where the record of dependencies in the form of *supports/2* unit clauses is put to use. Consequently, a new agenda is built up and rules which had been already executed become eligible to be executed again. For example, if the *span* is changed, the entry that has the value of *span* will be deleted. This will lead to everything that depended on span to be deleted, too. Additionally, when the new value of span is added, all the rules that have span in the *antecedents* will start re-firing and a new set of solutions will emerge on the blackboard, e.g., a new set of alternatives of structural systems will re-emerge. Figure 7.8a shows how the deletion of an entry from the blackboard causes the deletion of all the entries dependent on it using the *supports/2* clause. In the tree shown in Figure 7.8b, an entry is denoted by 'Nx' which is its *Tag No.* It is clear from this figure that the deletion of any entry causes the

```
stability       --->   stiffness
stiffness       --->   section size
section size    --->   dimensions
moment of inertia ---> section area
design strength --->   material
design strength --->   thickness of plates
```

Figure 7.7a Some of the Dependencies stored in the knowledge-base of DESCON

```
A  -->  B
B  -->  C
C  -->  D
D  -->  E
---------
---------
---------
X  -->  Y
```

Figure 7.7b Traversal of the chain of Dependent Entities

whole of that particular branch of the tree deleted from the blackboard at whose head the first entry to be deleted resides. If the top node 'N' is deleted, the whole tree is deleted and the blackboard becomes empty again. The system accomplishes this by traversing down all the branches under a node (entry) using the *supports/2* clause until it reaches a leaf node. One problem with the EPBS (12) as regards handling the amendment of an entry is that even the amendment of an entry is treated in much the same way as deleting one. The only difference is that the entry to be amended is replaced by the amended version but all the entries that depend on the entry to be amended are deleted in the same manner as if the original entry was deleted rather than amended. Although this is essential for maintaining the consistency of the blackboard (as explained in Section 2.7.2.4), a more sophisticated system would propagate the effect of amendment of an entry to its dependent entries rather than delete them. To overcome this problem, additional rules are included that explicitly manipulate the relationships between entities and infer the effects of change in one on the others based on the types of relationships between them as discussed in the previous section.

7.3.5 Knowledge Base

7.3.5.1 Introduction

The knowledge-base of DESCON is organised into the following three levels:-

(i) deep knowledge level;
(ii) heuristic level; and
(iii) general problem-solving level.

7.3.5.2 Deep Knowledge Level

This level consists of knowledge from text-books and other literature. There is hardly any expert knowledge involved in this prototype. The reason behind this is that in this experimental prototype, the aim was to solve a problem using fundamental knowledge rather than heuristics. In fact, one of the main reasons behind undertaking this experiment was to assess the difficulties involved in such an effort. There is only one knowledge module at this level as follows.

DEPENDENCE - this knowledge module consists of knowledge regarding the dependencies and relationships between different structural design entities, e.g., stiffness, stresses, moment, section sizes etc. The following is a rule from this module which is self-explanatory.

```
if      notnow[dependent,_,true]
then    true
to      add[dependent,(moment - section_mod),true]
        and[dependent,(moment - stress),true]
        and[dependent,(stress - material),true]
        and[dependent,(section_mod - moment_of_inertia),true]
        and[dependent,(moment_of_inertia - section_area),true]
        and[dependent,(moment_of_inertia - dist_neutral_axis),true]
        and[dependent,(section_area - section_dimension),true]
```

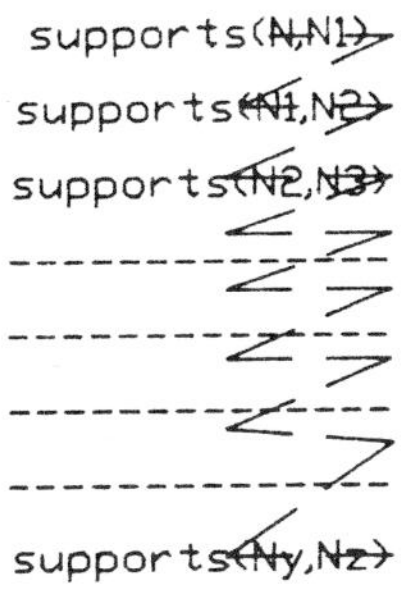

Figure 7.8a The dependent entry records on the blackboard and the propagation of the deletion of an entry

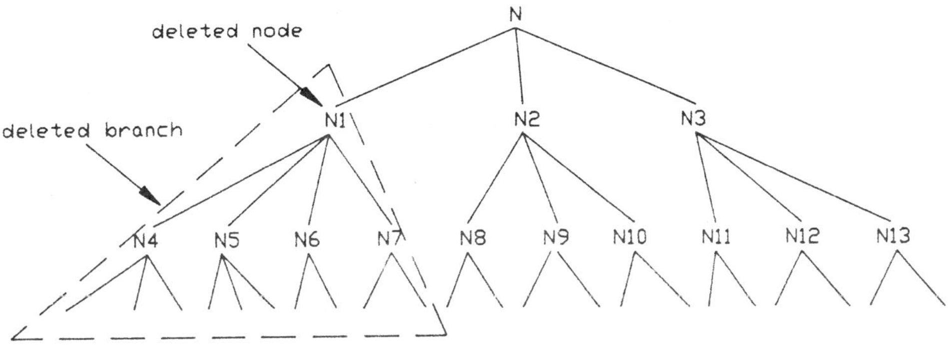

Figure 7.8b Deletion of a branch of the tree by deleting a node

```
        and[dependent,(dist_neutral_axis - section_dimension),true]
        and[dependent,(stiffness - section_size),true]
        and[dependent,(stability - stiffness),true]
est     depend(1).
```

This rule stores the fact that moment is dependent on stress and stress is dependent on material and so on.

7.3.5.3 Heuristics level

As the name suggests, this level contains rules of thumb or heuristics obtained from either literature or practicing engineers. There is only one module at this level as well, as follows.

EFFECTS - this knowledge-module incorporates heuristics regarding the effects of different design parameters on others. Following is a simple rule from this module:

```
if      notnow[effects,knowledge,true]
then    true
to      add[effects(eaves_haunch),load_carrying_capacity(increased),true]
        and[effects(purlins),lateral_restraints,true]
        and[effects(purlins),roof_support,true]
        and[effects(lateral_restraints),stability(increased),true]
        and[effects(ridge_haunch),deflection(decreased),true]
        and[effects(fixed_base),stability(increased),true]
        and[effects(fixed_base),frame_weight(decreased),true]
        and[effects(pinned_base),frame_weight(increased),true]
        and[effects,knowledge,true]
est     effects(1).
```

This rule stores the effects of some entities on the behaviour of the structure, e.g., the effect of eaves haunch is to increase the load carrying capacity.

7.3.5.4 Problem-Solving Level

This level contains knowledge regarding the actual solving of the problems based on the problem-solving strategy explained earlier in Section 7.3.2. There are two knowledge modules at this level as described below.

PROSOL - The knowledge incorporated at this level is specific to solving the particular types of problems that is intended to be solved using the prototype, i.e., finding alternative solutions when a particular solution is found infeasible due to some reason or the other. Following is a rule from this KM written in the blackboard shell syntax:

```
if      [primary_goal,find_alternatives_of(X),true]
then    find_supports_of(X,Y)
to      add[secondary_goal,find_dependence_of(Y),true]
est     prosol(1).
```

This rule states that if the primary goal is to find alternatives of X then backtrack to find the supports of X and add the secondary goal to find dependence of Y (which is the support of X found on backtracking) on the blackboard. 'find_supports_of(X,Y)' is a PROLOG procedure.

RELATIONS - This knowledge module incorporates general problem-solving capabilities for manipulating relationships between design entities. For example, if the relationship between two entities is *unconstrained directly_proportional*, it can infer the influence of a change in either of the entities on the other. It has been observed that most of the problem-solving activities in structural design revolve around finding such influences and inferring sensible results from them. Following is a rule from this module:

```
if      [value_of(A),B,true]
and     [relationship_between,(A,D,inversely_proportional(unconstrained)),true]
then    opposite(B,C)
to      add[value_of(D),C,true]
est     relations(1).
```

This rule states that if there is an *unconstrained inversely_proportional* relationship between two entities A and D and the value of A has undergone a change B (increased or decreased), then the value of D will also undergo a change C (increase or decrease as appropriate) which will be opposite in nature to the change in A (i.e., B).

7.4 Conclusions

1. The amount of knowledge required for the design of any real-life structure is vast and comes from various sources.
2. The different pieces of knowledge required to solve practical problems in structural design are often conflicting.
3. Some knowledge is even uncertain. This means that a particular piece of information could be true in some cases and untrue in some others. It is practically impossible to ascertain, in advance, the cases when a particular piece of information will be true and when it will not be.
4. A major proportion of the knowledge is based on the interpretation of results of different experimental research. In fact, codes of practice are a collection of rules derived from this interpretation. As such, this knowledge is dynamic and keeps changing with more and more research.
5. An 'intelligent' system for the design of structures should be sufficiently knowledgeable in different theoretical as well as practical aspects of design.
6. The theoretical aspects of design might require the capabilities to manipulate equations and infer sensibly from them.
7. Most of the knowledge required to solve real-life problems in structural design comes from the area of 'theory of structures' which deals with the analysis of structures.
8. Some problems can be solved by using simple heuristics acquired over the years of working experience.
9. Some problems require both heuristics as well as analytical knowledge about the analysis of structures.
10. One of the basic problems in detailed design is the proper interpretation of the provisions of codes of practice. This, in turn, depends largely upon the thorough understanding of theory of structures.
11. The most important capability that an 'intelligent' system should possess is a thorough understanding of the relevant areas of theory of structures. As this knowledge is highly mathematical, the system needs to have powerful

capabilities in mathematical manipulation.

12. A prototype developed on the lines of foregoing conclusions was seen to give encouraging results. The problem-solving methods used were backtracking and heuristic search.
13. The prototype was, however, found to be very slow. This was because the strategy adopted was to approach the problem from fundamental principles which requires extensive search and also a lot of other operations, particularly the manipulation of relationships between different entities.
14. A faster system would require faster hardware as well as other more 'intelligent' strategies.

References

1. **MacCallum, K.J.,** *Understanding Relationships in Marine Systems Design,* Proceeding of the First International Marine Systems Design Conference, London, pp. 1-9, 1982.

2. **Kumar, B. and Topping, B.H.V.,** *Issues in the Development of a Knowledge-Based System for the Detailed Design of Structures*, Artificial Intelligence in Engineering: Design, Ed. J.S. Gero, Computational Mechanics and Elsevier Publications, pp. 295-314, 1988.

3. **Lawson, B.,** *How Designers Think,* The Architectural Press Ltd., London, 1980.

4. **Sriram, D.,** *Knowledge-Based Approaches to Structural Design.* Ph.D. Dissertation, Department of Civil Engineering, Carnegie-Mellon University, U.S.A., 1986.

5. **Morris, L.J. and Randall, A.L.,** *Plastic Design,* Constrado Publications, Croydon, U.K., 1979.

6. **Morris, L.J.,** *A Commentary on Portal Frame Design,* The Structural Engineer, Volume 59A, No. 12, Journal of the Institution of Structural Engineers, London, December 1981, pages 394-404.

7. *Discussion - A commentary on portal frame design,* The Structural Engineer, Volume 61A, No. 6, Journal of the Institution of Structural Engineers, London, June 1983, pages 181-189.

8. *Correspondence - A commentary on portal frame design,* The Structural Engineer, Volume 61A, No.7, Journal of the Institution of Structural Engineers, London, July 1983, pages 212-221.

9. **Horne, M.R., Shakil-Khalil, H. and Akhtar, S.,** *The stability of tapered and haunched beams,* Proceedings of the Institution of Civil Engineers, Vol. 67, London, September 1979, pages 677-694.

10. **Rafiq, M.Y. and Macleod, I.A.,** *Logic Programming for Technical Design ,*

Department of Civil Engineering, University of Strathclyde, 1987.

11. **Rich, Elaine,** *Artificial Intelligence,* McGraw-Hill Book Company, pp 436, 1983.

12. **Chan, N. and Johnson, K.,** *Edinburgh Blackboard Shell User's Manual.* Artificial Intelligence Applications Institute, University of Edinburgh, 1987.

Chapter 8

Conclusions and Further Research

8.1 Introduction

The preceding chapters presented the components of the knowledge-based prototype, INDEX, in detail. The specific conclusions drawn from the development of each of these were listed in the respective chapters. This chapter will present the general and specific conclusions drawn from the whole work. Some suggestions on the limitations of INDEX and further work will also be given.

8.2 Some General Comments

In recent years, there has been considerable interest in knowledge-based applications to different areas of Civil Engineering. When this research commenced, Expert Systems technology appeared to some to answer all the questions raised in the preparation of computer software for Civil Engineering and, indeed, other areas. With scores of papers written on HI-RISE (1), the initial reaction was perhaps one of over-enthusiasm. Over the last few years, during the course of this work, the journals and conference proceedings have included many papers on "Expert Systems" to the extent that it becomes practically impossible to review all of them. However, the initial over-enthusiastic response seems to be dying down and researchers are starting to take a more realistic and mature view of this new technology. One of the general conclusions of this work as regards the potential of Expert System technology is that the initial expectations were too high. Most of the work reported so far has failed to convince that they are any different from the conventional programs. One of the important questions about the KBESs that must be addressed is that ''is it just another (possibly more convenient) 'means' to achieve the same 'ends' as with the conventional programs or could an entirely different 'end' be achieved ?''. The conclusion of this work is that most of the work reported has been just another way of programming and achieving essentially the same ends but with possibly more ease. That is not to say that that is all there is to the Expert System technology. It is thought that it was not necessary to use expert system shells in most of the works reported so far. It is also suggested that unless certain reasoning and problem-solving techniques from Artificial Intelligence are exploited, there is not much point in employing these tools. In fact, some 'learning' programs are written in BASIC (2). Even this project started with writing a Fortran program on plate girder design incorporating feedbacks from past designs and exhibiting some *crude* learning behaviour (Appendix VIII). This work has been an attempt at using the technology and other developments in Artificial Intelligence to achieve something different from the conventional programs. It must, however, be pointed out that throughout this work it was felt, and continuously strengthened, that the KBESs may only act as an assistant to the human designer and not replace him. As pointed out in Section 5.2.2.1, the human being's role still remains supreme and it is important to have this point clearly understood while making any attempt at developing an *intelligent* software.

8.3 Conclusions drawn from the development of the components of INDEX

8.3.1 Conclusions drawn from the development of ALTSEL

The ALTSEL module of INDEX is an illustration of a rule-based approach to developing design systems in a blackboard architecture environment. The Artificial Intelligent problem-solving methods used in ALTSEL have been described in Section 5.2.3. However, much of the problem-solving in ALTSEL is undertaken by utilising the inherent features of the blackboard architecture as implemented in the Edinburgh Prolog Blackboard Shell. ALTSEL may be seen to be an application of a new programming technique to design. The most important conclusion drawn from the development of ALTSEL is the ease of encoding pieces of information in rule-form using an expert system development tool which would have been quite complicated in other procedural languages such as FORTRAN. Also, the execution of rules in the required order suitable to the domain could be achieved with minimum effort.

8.3.2 Conclusions drawn from the development of DETDEX

The development of DETDEX proves the utility of knowledge-based technology to automated standards processing with certainty. The ease of representing a standard as facts and manipulating them with general rules proves the edge this approach has over other representations of standards, e.g., rules, decision tables, etc. The most important feature of DETDEX is its generic standards processing capabilities. Factual representation of standards was found to have an edge over the production rule representation not only in terms of ease but also in terms of savings in computer memory. The other important advantage of this representation is that the code-dependent and independent information are completely separate. The rule-based representation suffers from the drawback that each rule not only contains the information from a standard but also does the processing too. Thus, any update to the standard can only be incorporated in the system by updating the whole program whereas in a factual representation, only the relevant facts have to be changed and the processor remains untouched. The task of updating the facts is straightforward and does not require any computing skills at all.

8.3.3 Conclusions drawn from the development of DESCON

Out of all the modules developed so far, DESCON has the most potential. The problem-solving techniques used by DESCON were found to be sufficiently powerful and a comprehensive knowledge-base could enhance its performance. The network model as proposed by MacCallum (3) was found to be quite powerful. However, to utilise it effectively, it was felt that it needs to be supplemented with different problem-solving methods from AI such as Dependency-directed backtracking in DESCON. The explicit representation of dependencies and relationships between entities proved to be a powerful technique for resolving violation of constraints in design. This feature of explicit representation of dependent entities in DESCON is certainly seen to be a step forward from the conventional software in which these dependencies and relationships are implicit in the procedures encoded in the program. Such an implicit representation cannot be used to resolve conflicts between constraints or constraint violation so fundamental to design. The *numerical* manipulation of relationships between entities was demonstrated by DESIGNER (3) and, more recently, RELATOR (4). DESCON has demonstrated the use of the same network model supplemented by other AI techniques in resolving constraint violation and change in specifications by *symbolic* manipulation of relationships between dependent entities.

8.3.4 General Conclusions

It must be pointed out that this work has been an attempt at undertaking some experiments with the applications of some of the developments in Artificial Intelligence to structural design and not developing a fully working expert system. Some important general conclusions drawn from these experiments are discussed below.

8.3.4.1 Knowledge Representation

Since the EPBS is a forward-chaining rule-based shell, most of the knowledge represented in INDEX is as production rules. The biggest advantage of using a production rule representation was the ease of doing so. However, on many occasions, its limitations proved quite an obstacle. The most important limitation was its inability to represent a declarative piece of information. For example, if a description of a certain object was to be stored for use at any stage in the solution process, it could not be done easily. This leads to its other drawback of not being able to manipulate a knowledge-base by a separate set of *control* rules. In a KBES, all the knowledge should be stored before its processing is undertaken as and when required. In a rule-based system, this seems difficult to achieve because the rules are *self-processing*. It may not always be a limitation when a *direct inferencing* is all that is undertaken by the system and there is no search or manipulation of a static piece of knowledge required. In fact, ALTSEL works mostly as a *direct inferencing* system. This limitation was experienced while developing DETDEX and DESCON when there was a clear need for having a separate set of *control* rules to manipulate a static piece of knowledge. This raises the issue that **there is a need for a multi-formalism environment** in which the knowledge-base may (possibly) be represented in any object-orientated language and the *control* rules may be represented as production rules. Multi-formalism based knowledge-based toolkits such as KEE and LOOPS (5) seem suitable for this purpose.

8.3.4.2 Architecture

The *blackboard* architecture adopted for INDEX proved appropriate for the domain. In fact, the very nature of a design problem fits exactly with the features of the blackboard architecture. It is essential to have a number of interacting sources of knowledge in design and the blackboard architecture supports just that.

8.3.4.3 Use of an Expert System Shell

Using the Expert System shell EPBS helped speed up the implementation to a great extent. It is felt that using a language rather than a shell would probably be better as a language may provide more flexibility than a shell but the significant question is the time required to do so. It is worthwhile to spend more time researching the philosophical issues involved rather than programming. The shell used, however, must be flexible enough to support the requirements of the project. The EPBS proved to be quite a powerful research tool for the reasons discussed in Section 2.7.2.5. It did have its limitations, e.g., inability to generate explanations, lack of a facility for the inheritance of properties, but the provision to change to Prolog programming as and when required helped solve most of these limitations.

8.3.4.4 Interface with Algorithmic Programs

In a domain such as structural design, it was felt imperative to have the facility to interact with algorithmic programs. The structural analysis programs are indispensible for any structural design system and, thus, having an interface with these programs is a must.

8.3.4.5 Problem-solving Techniques

As regards problem-solving techniques required, it was felt that it was essential to have more than one problem-solving technique. It is thought that no design system is possible without the ability to employ different techniques in different stages of design. For example, for some parts (as in ALTSEL) *direct inferencing* rules may be adequate but for others (as in DESCON) DDB may have to be used.

8.4 Suggestions for Further Research

It must be pointed out that this work has been an attempt at undertaking some experiments with the applications of some of the developments in Artificial Intelligence to structural design and not developing a fully working expert system. As with any research work, there are a number of extensions to INDEX that could be undertaken in future. Some of the general ones are listed below:

1. calibration of the system against real-life problems;
2. extension of the knowledge-bases of the modules;
3. implementation of the other modules of INDEX;
4. incorporation of explanation facilities;
5. re-implementing the system in a multi-formalism environment and compare with the current implementation.

In the following sections, specific extensions to the modules developed in this work will be suggested.

8.4.1 Extensions to ALTSEL

The problem-solving method used in ALTSEL is the *derivation approach* discussed in Section 5.2.3. The reasons behind adopting this approach were also given in the same section. It certainly proved easier to implement but the biggest handicap is the inability to generate explanations. The only explanation ALTSEL can generate is to inform the user about the *antecedents* of the rule that forces the conclusion in question. It is thought that a *formation approach* (Section 5.2.3) will help in generating comprehensive explanations.

8.4.2 Extensions to DETDEX

In DETDEX, the standard provisions have been represented as facts. The different types of facts have been discussed in Chapter 6. In the performance facts, one of its parts is the involved data-item. In the current representation scheme, this data-item has to be a uniquely defined data-item and thus a general data-item (e.g., deflection) cannot be handled by this scheme. In the current scheme, if a provision says that the deflection at certain points should not exceed a certain value, the deflection at all the concerned points will have to be represented separately, thus, producing multiple facts for the same provision. Quite clearly, this is a limitation of the representation scheme suggested and needs further work.

At the moment, DETDEX suns very slowly. The reason for this is quite evident. DETDEX has to find all the expressions for all the derived data-items before it starts processing them. If it finds a derived data-item in an expression, it has to again search for the expression for that data-item and so on. Once it has found all the expressions for all the data-items, it starts processing backwards finally reaching the original expression to be evaluated. It has to do this for all the data-items concerned in a particular problem. In case of a general conformance checking for a structure, as the number of involved data-items will be large, DETDEX has to do even more work and hence it is even slower. One improvement in this regard could be to re-implement

DETDEX on a parallel processor. The nature of the problem seems quite ideally suited to parallel processing. All the inputs to the system are read in right at the start and the standards processor has to follow similar steps for all the calculations, e.g., for finding the principal involved data-item, finding its expression, finding if there are any derived data-items in the expression and if there are, then finding their expressions and so on and finally processing them. All this work has to be repeated for all the principal data-items for all the concerned provisions. Quite clearly, this can be done in parallel as all the inputs are already present right in the beginning and the same program has to be executed repeatedly. This type of parallelism is called event parallelism (6).

8.4.3 Extensions to DESCON

The most important extension to DESCON can be in the representation of the input of soft constraints. At the moment, it has a very crude representation in the sense that any removal of some parts of the structure are input as a *physical infeasibility* problem and the violation of constraints are simply posted as an unsatisfied condition on the blackboard. However, if a more vague constraint has to be introduced, e.g., change the location of the building, DESCON cannot handle that. The other area is the incorporation of more heuristics to guess at the amount of increase or decrease in a certain entity if it is suggested to increase or decrease some entity's value.

The other important extension to DESCON would be to enhance its capabilities to handle situations where more than one entity is varying simultaneously, i.e., multivariate analysis. At the moment, the relationships handled by DESCON are too simplistic. The relationships between all the entities may not be as simple as *directly proportional* or *inversely proportional*. The incorporation of such features will obviously also require the strengthening of DESCON's knowledge-base, too. The abstraction hierarchy given in Figure 7.5 may be helpful in extending DESCON's knowledge-base and problem-solving capabilities.

The knowledge-base of INDEX's other modules need to be extended in order to realise the full use of DESCON. At the moment, in some cases, the knowledge-base of other modules are not comprehensive enough to execute the suggestions given by DESCON and the system stops at that point. This is not seen as a major limitation as overcoming this problem may simply require expanding the knowledge-base of INDEX and not much fundamental research may be required.

References

1. **Maher, M.L. and Fenves, S.J.**, "*HI-RISE: An Expert System for the Preliminary Design of High Rise Buildings*", Report No. R-85-146, Department of Civil Engineering, Carnegie-Mellon University, Pittsburgh, U.S.A., 1985.

2. **Naylor, Chris**, "*Build your own Expert Systems*", Sigma Press, 1986.

3. **MacCallum, K.J.**, "*Creative Ship Design by Computer*", Computer Applications in the Automation of Shipyard Operation and Ship Design IV, Eds. D.F. Roger et. al., North-Holland Publishing Company, pp. 55-62, 1982.

4. **Rafiq, M.Y. and Macleod, I.A.**, *Logic Programming for Technical Design*, Department of Civil Engineering, University of Strathclyde, 1987.

5. **Lurent, J., et. al.**, "*Comparative Evaluation of Three Expert Systems Development Tools: KEE, Knowledge Craft and ART*", The Knowledge Engineering Review, Vol. 1, No. 4, pp. 18-29, 1986.

6. **Bowler, K.C.** *et.al., An Introduction to Occam2 Programming,* Chartwell-Bratt, 1987.

Appendix I

Interface between Fortran and Prolog

I.1. Introduction

This appendix describes two approaches to the development of an interface between PROLOG and FORTRAN. These are :

1. The File Approach and
2. The C Interface Approach.

The design and implementation of the C interface approach will be given and compared with the file approach under the UNIX operating system.

I.2. The Interface

I.2.1 Introduction

The file approach - this is a simple way of developing an interface. In this approach, the knowledge-based component and FORTRAN communicate via files. Some implementations of PROLOG (e.g., Quintus-PROLOG, Edinburgh PROLOG and C-PROLOG) allow for calling the operating system and executing any system command from PROLOG. This facility is the key to this method. The knowledge-based component stores all the input data for the FORTRAN program in a file. The FORTRAN program is invoked by a system call from PROLOG. Similarly, the FORTRAN program stores its output on a file which is read from PROLOG. This approach is quite straightforward. However, it does not provide a fully integrated environment.

The C interface approach - with this approach, the communication between PROLOG and FORTRAN is through C functions as the intermediary. Some implementations of PROLOG (e.g., Quintus-PROLOG, Edinburgh PROLOG) allow for loading of pre-compiled object codes. On the other hand, the Berkley Unix implementation of FORTRAN provides for calling pre-compiled FORTRAN function from C and vice-versa. Thus, the strategy adopted for INDEX was to call C from PROLOG and then FORTRAN from C. To accomplish this, both the C and FORTRAN functions are compiled and linked together in one file. This linked file is, then, loaded in Edinburgh PROLOG. Since the compiled FORTRAN and C functions are linked together, both share common data storage (see Figure I.1). Thus, FORTRAN functions have direct access to any data passed from PROLOG to C. Similarly, any output data from FORTRAN is directly accessed and sent to PROLOG by C.

Pre-compiled codes may be loaded in Edinburgh PROLOG (1) by the load/3 predicate, where 3 stands for the number of its arguments, as given below :

```
load(ListofPredSpec,ObjectFiles,Libraries).
```

A simple example for loading pre-compiled C functions is :

```
load([pred/1=cfunc1, pred/2=cfunc2],'tmp.o','-lm -termcap').
```

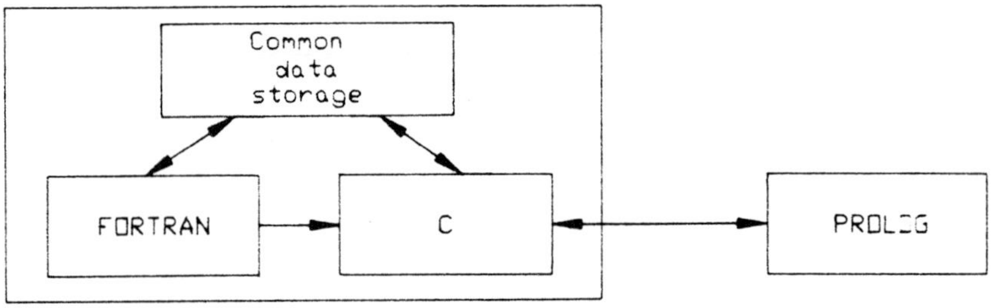

Figure I.1 The interface

$$[A] = \begin{bmatrix} A11 & A12 & A13 \\ A21 & A22 & A23 \\ A31 & A32 & A33 \end{bmatrix} \Rightarrow \big[[A11,A12,A13], [A21,A22,A23], [A31,A32,A33]\big]$$

Figure I.2 Representation of arrays in PROLOG

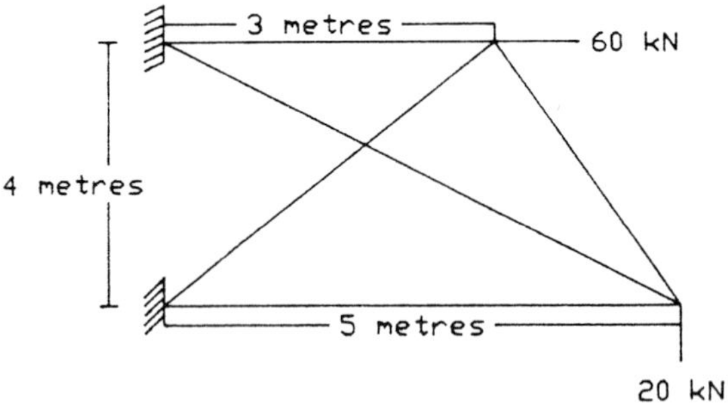

Figure I.3 The test structure

The first argument ListofPredSpec is a list of predicate specifications, each of which specifies a PROLOG predicate and its arity that is to be associated with a C function. In the example given above two C functions @I[cfunc1] and @I[cfunc2] are to be loaded. These functions are to be called from PROLOG by pred/1 and pred/2 respectively.

In the present version, only simple data types, i.e., integers, floats and atoms, may be passed between C and PROLOG. There are three C functions provided for the C code to access the arguments passed to it from PROLOG. They are :

getint(),getfloat() and getatom()

Another three C functions are provided for the C code to pass back the values to PROLOG. They are :

putint(), putfloat() and putatom()

A brief description of these functions (1) is as follows :

getint(ArgNo, AddrOfInt) - @I[ArgNo] is a positive integer specifying which PROLOG argument is to be accessed. @I[AddrOfInt] is the address of a C integer variable. getint(ArgNo,AddrOfInt) assigns the value of the @I[ArgNo]th argument of the PROLOG call to the @I[AddrOfInt] variable and returns the value 1. If the corresponding PROLOG argument is not an integer the assignment operation is skipped and the function returns the value 0.

getfloat(ArgNo,AddrOfFloat) - This function is similar to getint(), but is used for accessing floating point numbers rather than integers. @I[AddrOfFloat] should be the address of a C variable declared as double.

getatom(ArgNo,AddrOfPointerToString) - This function is for accessing a PROLOG atom. The meaning of ArgNo is the same as for getint() and getfloat(). @I[AddrOfPointerToString] should be the address of a pointer to a string of characters. The function makes the string pointer point to a string of characters representing the PROLOG atom. If the specified argument is not an atom, the function returns the value 0.

putint(ArgNo,IntValue) - @I[ArgNo] is a positive integer specifying which argument in PROLOG is to be unified with the integer value @I[IntValue]. If the unification succeeds the function returns the value 1, otherwise the value 0.

putfloat(ArgNo, FloatValue) - This is similar to putint() except that it is used for floating point numbers.

putatom(ArgNo,PointerToString) - This function is similar to putint() except that the second argument should be a pointer to a string. This function creates a PROLOG atom from the string and unifies it with the specified argument. If the unification succeeds the function returns the value 1, otherwise the value 0.

I.2.2 Accessing arrays

The communication of simple variables does not prove any problem but accessing arrays requires further consideration. The arrays are represented as lists in PROLOG. One-dimensional arrays are represented as single lists whereas multi-dimensional ones are represented as lists containing sub-lists as shown in Figure I.2. Each sub-list in

Figure I.2 represents a row of the equivalent C two-dimensional array. The array that is passed back from C is also a list consisting of sub-lists. The methods of accessing these arrays in the two methods of interfacing are described below.

The file approach - the elements of the lists representing the arrays in PROLOG are written to a file which is then read by the FORTRAN program. The FORTRAN program writes its output on a file. The output arrays are then read and formed into lists by PROLOG. Each element of the arrays is passed back to PROLOG one by one and the corresponding list keeps getting constructed in PROLOG using a simple PROLOG procedure .

The C interface approach - the elements of the lists representing the arrays in PROLOG are passed one by one recursively to FORTRAN via C. Similarly, the elements of the output arrays from FORTRAN are passed back one by one and the corresponding list gets constructed recursively in PROLOG.

Considering the mapping of arrays between C and FORTRAN, the most important requirement is that the dimensions of the arrays have to be altered. This is because FORTRAN 77 arrays are stored in column-major order whereas C arrays are stored in row-major order. Thus, the column dimension of a FORTRAN 77 array will become the row dimension of the equivalent C array and the row dimension of FORTRAN 77 array will become the column dimension of the equivalent C array. For example, an array A(3,2) in FORTRAN will have to be represented as A[2][3] in C and will be the transpose of the actual array required in FORTRAN. Another important point is that the first element of a C array always has a subscript zero whereas FORTRAN 77 array elements always begin with a subscript of one.

I.2.3 Performance measurement

An ideal performance measurement for our purposes would have been to compare the cpu times taken in running an analysis program written in PROLOG and a similar program written in FORTRAN using the interface. Such a test would have ascertained whether rewriting the analysis program in PROLOG was worthwhile or not. If a PROLOG program ran faster than the FORTRAN one running with either of the two methods of interfacing, it would be worthwhile to rewrite the FORTRAN programs in PROLOG. But, such a decision would mean losing all the investment made in writing FORTRAN programs and making new investments in rewriting the same programs in PROLOG. This is a crucial decision as rewriting the programs in PROLOG would need quite a lot of effort and time, which might not be desirable. It can be very safely said that a PROLOG program to perform the analysis would be slower than an equivalent FORTRAN program. In general, PROLOG is slow in performing numerical operations. Here, an evaluation of the performance of the two approaches is presented by comparing the times in running a FORTRAN structural analysis program in the following different ways:

1. running the FORTRAN program without any interface, i.e., reading the data directly from a file and writing the output on another file,
2. running the FORTRAN program using the 'C interface approach' and
3. running the FORTRAN program using the 'file approach'.

These tests were carried out using the same set of data. The test structure and loads are given in Figure I.3. This gives an indication of relative overheads of using the two approaches to interfacing.

The results of these tests are given in the following table.

Test No.	Test Description	Average cpu time of three runs (in centiseconds)
1	the FORTRAN program running without any interface	10
2	the FORTRAN program running using the C interface	13
3	the FORTRAN program running using the file approach interface	35

It is clear from these results that the C interface approach is faster than the file approach. It is very marginally slower than the FORTRAN program running by itself without using any interface. On the other hand, the file approach is considerably slower. Since the computing time for analysis is the same in both cases, the time difference between the two approaches is equal to the difference in time taken by each in transferring the arrays and other data between FORTRAN and PROLOG.

To obtain some idea of the time spent in passing and returning arrays of different sizes by each approach, another test was undertaken. The results of these tests are shown in the following table.

Array size	Average cpu time of three runs (centiseconds.)	
	The C interface approach	The file approach
10x10	15	69
25x10	38	165
30x10	45	205
50x10	74	336
100x10	146	675

These results explain the results shown in the table given earlier. The following table actually gives an indication of the relative overheads involved in using the two methods of interfacing.

Based on the results of the two tables given above, it is clear that the C interface approach is almost three to four times faster than the file approach, on an average. However, for the test problem tried, the difference was only around 25 centiseconds on an average. Thus, it may be concluded that the choice between the two approaches will become crucial only when the number and sizes of the arrays to be used by the FORTRAN program are quite large. For example, if a program uses 100 arrays of 1000 elements each, the difference in time in just passing these arrays would be almost 7 minutes. Although, this difference would not be significant for a program that runs for hours, it is quite considerable. On the other hand, for graphics and other user interface facilities, time is of utmost importance and the obvious choice in that case would be the faster approach even if the time difference is minimal.

I.3 Limitations of the two approaches

The C interface approach - one of the most important limitations of this approach is that the FORTRAN program cannot be used straightaway as it is originally written. Minor modifications have to be made to it in order to load them in PROLOG. They are as follows:

1. to change the FORTRAN program into a function/subroutine, which means removing the main program and changing it into a function which calls other functions/subroutines and
2. to remove the input and output statements since it is difficult to open input/output channels in FORTRAN while inside PROLOG.

These changes did not prove too difficult to us since the routines were written by ourselves (Kumar and Topping) only. However, there could be problems on this account while using a commercial package due to the source being inaccessible to the user. This could be seen as a serious limitation of this interface.

In terms of speed, the above-mentioned results clearly show that it is faster than the file approach. Another merit of this approach is that it provides us with a fully integrated environment in the sense that all the components, i.e., PROLOG and FORTRAN, of the expert system are interlinked and the whole system behaves as one single unit.

The file approach - this method does not require any modifications to be made to the FORTRAN program. However, in terms of speed, this approach is considerably slower. Another important disadvantage of this method is that it does not provide a fully integrated environment.

References

1. Artificial Intelligence Applications Institute, University of Edinburgh, "Edinburgh PROLOG (The New Implementation) User's Manual, Version 1.4", 1986.

Appendix II

The following is a sample run of a simple rule-based standards processing prototype based on some provisions from Section 5 of BS5950 also illustrating some explanation capabilities in the second run.

%%% User inputs and comments are in bold typeface

Edinburgh Prolog, version 1.5.01 (14 August 1987)^M
AI Applications Institute, University of Edinburgh^M ^

| ?- **run.**

What type of design check do you want to carry out?
  1 elastic
  2 plastic
Type in a number followed by a full stop: **2.**

Is the load predominately static?
  1 yes
  2 no
Type in a number followed by a full stop: **1.**

Does the steel used comply with BS 4360?
  1 yes
  2 no
Type in a number followed by a full stop: **1.**

What is the thickness of the plates to be used?
  1 16
  2 25
  3 40
  4 63
  5 100
Type in a number followed by a full stop: **1.**

What is the grade of the steel to be used?
  1 43
  2 50
  3 55
Type in a number followed by a full stop: **1.**

****
Based on Table 6, the design-strength of the steel is
275 N/sq.mm.
****

Are the flanges symmetrical ?
1 yes
2 no
Type in a number followed by a full stop: **1.**

What is the thickness of the web (in mm.) ? **12.**

What is the depth of the web (in mm.) ? **300.**

What is the width of the flange (in mm.) ? **40.**

What is the thickness of the flanges (in mm.) ? **12.**

****
The area of the section is 4560
****

****
The torsional constant 'J' for the section is 47280.0
****

****
The distance between shear centres of the flanges is
312.0
****

What is the axis of symmetry of the section ?
1 major
2 minor
Type in a number followed by a full stop: **1.**

What is the maximum distance between a hinge restraint (if any) and an adjacent restraint (in mm.) ?
1212

What is the height of the column (in mm.) ? **1222.**

What is the deflection at the top of the column under notional horizontal loading (in mm.) ? **12.**

```
Test failed:

************************************************************
As per Section 5.5.3.2 :
Deflection should not exceed h/1000, where h is the height
of the column.
************************************************************


What is the average compressive stress due to axial load ?      122.


%%%% Another run illustrating some explanation capabilities


| ?- run.


What type of design check do you want to carry out?
        1     elastic
        2     plastic
Type in a number followed by a full stop: why.


This program checks the BS5950 conditions for plastic design only


What type of design check do you want to carry out?
        1     elastic
        2     plastic
Type in a number followed by a full stop: 1.


************************************************************
As per Section 5.2 :
Elastic analysis should be carried out under factored loads.
************************************************************


[Execution aborted]

| ?- start.

yes
| ?- run.


What type of design check do you want to carry out?
        1     elastic
        2     plastic
```

Type in a number followed by a full stop: **2.**

Is the load predominately static?
  1 yes
  2 no
Type in a number followed by a full stop: **why.**

do not know

Is the load predominately static?
  1 yes
  2 no
Type in a number followed by a full stop: **1.**

Does the steel used comply with BS 4360?
  1 yes
  2 no
Type in a number followed by a full stop: **1.**

What is the thickness of the plates to be used?
  1 16
  2 25
  3 40
  4 63
  5 100
Type in a number followed by a full stop: **why.**

It is needed to consult Table 6 to get the strength of steel.

What is the thickness of the plates to be used?
  1 16
  2 25
  3 40
  4 63
  5 100
Type in a number followed by a full stop: ^C

Prolog terminated

Appendix III

The following is a sample run of the ALTSEL module as a standalone prototype. In the later part of the appendix, some explanation capabilities are also illustrated.

%%% User inputs and comments are in bold typeface

Edinburgh Prolog, version 1.5.01 (14 August 1987)
AI Applications Institute, University of Edinburgh

| ?- start,run.

What is the average vertical load (in kN/sq.m)? (The vertical
load is assumed to be uniformly distributed over the whole span)

3.

What is the eaves height (in metres)?

7.6.

Is there any horizontal load acting on the frame ?
1 yes
2 no
Type in a number followed by a full stop:**2.**

What is the span of the building ? (in metres)

55.

What is the pitch allowed for the portal frame (if possible) ? (in radians)

0.33.

Are internal columns allowed in the building ?
1 yes
2 no
Type in a number followed by a full stop:**2.**

What is the type of bases ?
1 fixed
2 pinned
Type in a number followed by a full stop:**1.**

Are cranes to be present in the building ?
1 yes
2 no
Type in a number followed by a full stop:**2.**

****
Trying to find which are the possible lateral load systems :-
****

single-span portal lateral load system is possible

tied portal lateral load system is possible

Multi bay latticed girder lateral load system is possible

The following alternatives of multi-bay portal are also feasible :-
1. Heavy portals at convenient spacings with ridge and valley beams carrying intermediate rafters,
2. Series of cranked rafters carried on eaves stanchion and valley beams.

The frame spacing for the span under consideration should be in the range of 10 to 12 metres and lattice purlins would prove economical. However, in this case, an intermediate frame becomes a necessity between two main frames. Type in a desired value followed by a full-stop. **11.**

****
The following alternatives can be considered for
the sides :-
****

One alternative is to just have side rails attached to the side stanchions.

Another alternative could be have side bracings between the stanchions.

****
The following alternatives can be considered for the
side cladding :-
****

One alternative for side cladding is to have plastic coated sheeting all over.

Another alternative for the side cladding is to have brickwork in one of the following ways :-

1.supported from the structure both vertically and horizontally,
2.supported only horizontally,
3.self_supporting both vertically and horizontally,
4.self_supporting and also supporting some elements such as the ends of purlins at the gables.

Another alternative for the side cladding is to have precast or cast-in-situ concrete wall all over.

****
The following alternatives can be considered for
the roof system :-
****

One alternative for the roofing system is to have cladding simply over purlins.

Another alternative for the roofing system is to have bracings between the rafters of the supporting frame.

****
Following are the approximate section sizes for the
different alternatives of feasible lateral load systems :-
****

Following are the feasible sections for the single span portal alternative :-

533x165UB@73
Zp provided = 1776

The following are the feasible sections for the tied portal rafters and columns based on aproximate analysis :-

381x152UB@52
Zp provided = 959.0

Following is the dimensions for the tie based on approximate analysis :-

60x60x10 angle or a rod of 36mm. dia.

****
The following design constraints should be considered
in the detailed design stage :-
****

The following things should be considered in the detailed design stage of single span portal alternative :-

1.pitch should be kept low because greater slope will give rise to greater spread at knees which can cause problems with cladding,

2.horizontal thrusts should be carefully examined and the foundation designed accordingly,

3.haunch should be provided at the eaves and the ridge should be deepened because the maximium bending moment will occur at the knees.

The following constraints should be taken into account in the detailed design stage of multi-bay portals with cranked rafters :-

a)spread under load could be critical and should be examined carefully.

yes
| ?- **show_details_of(single_span_portal).**

%%% type of section of the columns
section_type

column(ub)

%%% type of section of the rafters

section_type

rafter(ub)

%%% section modulus provided

ssp_zp_provided

1776

%%% section provided

ssp_feas_sec

533x165UB@73

```
yes
| ?- how(single_span_portal).
```

The reasons for single_span_portal being one of the feasible alternatives is beacause of the following entries being true:

```
Entry 276 as     Tag, Status, Cf
                 Index,
                 Fact is :-

276  in  true
problem
int_stanch(no)


Entry 271 as     Tag, Status, Cf
                 Index,
                 Fact is :-

271  in  true
problem
span(55)


yes
| ?- rule(single_span_portal).


Rule 25 is :-

    if   [problem,span(_125510),true]
         and [problem,int_stanch(no),true]
         and holds _125510=<60
    then  output_message(single-span portal lateral load system is possible)
    to    add [synthesis,lateral_load_sys(single_span_portal),true]
    est   synthesis(2)


yes
| ?- ^D
Prolog terminated
```

Appendix IV

The following is a listing of a portion of the rule-sequencing strategy of INDEX.

```
%%%% Abolish the default strategy provided in the shell

:- abolish(less_rating,2).
:- abolish(rating,2).

%%%% Top-level rule

rating(A,A).

less_rating(A,B) :- less(A,B),!.
less_rating(A,C) :- less(A,B),less_rating(B,C).

%%%% Specific rules pertaining to est values of rules inside different
%%%% knowledge modules

less_rating(top_control(_),_) :- !.
less_rating(top_control(_),control(_)) :- !.
less_rating(control(_),_) :- !.
less(control(_),synthesis(_)) :- !.
less(synthesis(_),crane(_)) :- !.
less(crane(_),fr_sp(_)) :- !.
less(fr_sp(_),side(_)) :- !.
less(side(_),roof(_)) :- !.
less(roof(_),preana(_)) :- !.
less(preana(_),predes(_)) :- !.
less(predes(_),eco(_)) :- !.
less(eco(_),des(_)) :- !.
less(des(_),evaluate(_)) :- !.
less(evaluate(_),data(_)) :- !.
less(data(_),prel_out(_)) :- !.
less(prel_out(_),design_check(_)) :- !.
less(design_check(_),cons(_)) :- !.


less(top_control(X),top_control(Y)) :- integer(X),integer(Y),X < Y, !.
less(control(X),control(Y)) :- integer(X),integer(Y),X < Y, !.
less(synthesis(X),synthesis(Y)) :-integer(X),integer(Y), X < Y, !.
less(crane(X),crane(Y)) :- integer(X),integer(Y),X < Y, !.
less(fr_sp(X),fr_sp(Y)) :-integer(X),integer(Y), X < Y, !.
less(side(X),side(Y)) :- integer(X),integer(Y),X < Y, !.
less(roof(X),roof(Y)) :- integer(X),integer(Y),X < Y, !.
less(preana(X),preana(Y)) :- integer(X),integer(Y),X < Y, !.
less(des(X),des(Y)) :- integer(X),integer(Y),X < Y, !.
less(predes(X),predes(Y)) :-integer(X),integer(Y), X < Y, !.
less(eco(X),eco(Y)) :- integer(X),integer(Y),X < Y, !.
less(evaluate(X),evaluate(Y)) :- integer(X),integer(Y),X < Y, !.
```

```
less(data(X),data(Y)) :- integer(X),integer(Y),X < Y, !.
less(prel_out(X),prel_out(Y)) :- integer(X),integer(Y),X < Y, !.
less(design_check(X),design_check(Y)):-integer(X),integer(Y),X < Y, !.
less(cons(X),cons(Y)):-integer(X),integer(Y),X < Y, !.
```

Appendix V

This appendix contains listings of runs of the DETDEX module as a standalone prototype. Three runs are given to illustrate the different capabilities of DETDEX. Section V.1 includes a listing of a run of the standards processing part of DETDEX which is called STAPRO. Section V.2 is also a listing of a run of STAPRO illustrating its backtracking facility in case of non-conformity to a provision of the code. Provisions in STAPRO are taken from the Section V of BS5950 which concerns the design of continuous construction. The inputs are such as to keep the number of applicable clauses at the very minimum. The main purpose of the run is to show how STAPRO works. Section V.3 includes a run of that part of DETDEX which can advise on the general considerations in the detailed design stage.

The user inputs and comments are in bold typeface. The units for loads (e.g., alo) are Newtons and for lengths (e.g., dw, tw, wf, tf, lm) are millimetres.

V.1 A sample run of STAPRO

Edinburgh Prolog, version 1.5.01 (14 August 1987)
AI Applications Institute, University of Edinburgh

| ?- **start,run.**

What is the problem?
1 design
2 design_check
Type in a number followed by a full stop:**2.**

What type of design do you want to carry out?
1 elastic
2 plastic
Type in a number followed by a full stop:**2.**

Which of the following do you want to carry out ?
1 conformance
2 determination
Type in a number followed by a full stop:**1.**

What type of conformance checking is required ?
1 data item
2 provision
3 general conformance
Type in a number followed by a full stop:**3.**

What is the primary object involved ?
1 structure
2 component
3 secondary_component
Type in a number followed by a full stop:**1.**

What is the structure involved ?
1 portal_frame
2 truss_and_columns
3 beam_and_columns
Type in a number followed by a full stop:**1.**

Is the following a limit state ?

rafter ,(plastic,,(buckling,,(yield,whole)))

yes.

Is the following a limit state ?

rafter ,(plastic,,(buckling,,(yield,local)))

no.

Is the following a limit state ?

portal_frame(rafter) ,(plastic,,(stability,whole))

no.

Is the following a limit state ?

portal_frame(rafter) ,(plastic,,(snapthrough_stability,whole))

no.

Is the following a limit state ?

portal_frame(column) ,(plastic,,(stability,whole))

no.

Is eaves haunch present ?
1 yes

2 no

Type in a number followed by a full stop:**2.**

Is apex haunch present ?

1 yes

2 no

Type in a number followed by a full stop:**2.**

Is the following a limit state ?

portal_frame ,(plastic,,(sway_stability,whole))

no.

==> Checking the applicability criteria of the concerned clauses

==> Checking the applicability criteria of clause 5.3.7

==> Any holes on either side of the hinge upto a length equal to the depth of any member either punched or drilled 2mm in diameter undersize and reamed :true or false **true.**

==> Checking the applicability criteria of clause 5.3.6

==> No hinge stiffener located at more than half the depth of the section on either side :true or false **true.**

==> Checking the applicability criteria of clause 5.3.5

==> No applicability criteria for clause 5.3.5 found.

==> Checking the applicability criteria of clause 5.3.4

==> No applicability criteria for clause 5.3.4 found.

==> Checking the applicability criteria of clause 5.3.3

==> Steel complies to BS4360 :true or false **true.**

==> Checking the applicability criteria of clause 5.3.2

==> Load predominantly static :true or false **true.**

==> Please input the actual value of lm for the column: **350.**

==> Please input the actual value of lm for the rafter: **300.**

==> The following queries relate to the parameters of column

==> Please input the value of alo: **1500.**

==> Please input the value of dw: **70.**

==> Please input the value of tw: **10.**

==> Please input the value of wf: **45.**
==> Please input the value of tf: **10.**

==> The following queries relate to the parameters of rafter
==> Please input the value of alo: **1650.**
==> Please input the value of dw: **explain.**

You mean explain dw!

==> dw is depth of web

==> Please input the value of dw: **70.**
==> Please input the value of tw: **10.**
==> Please input the value of wf: **45.**
==> Please input the value of tf: **10.**

==> The following queries relate to the parameters of column
for the determination of lm
==> Value of rry already present
==> Value of fc already present
==> Value of py already present
==> Value of x already present

==> The following queries relate to the parameters of column
for the determination of out_b_t_ratio
==> Value of py already present

==> The following queries relate to the parameters of rafter
for the determination of lm
==> Value of rry already present
==> Value of fc already present
==> Value of py already present
==> Value of x already present

==> The following queries relate to the parameters of rafter
for the determination of out_b_t_ratio
==> Value of py already present

%%% Following are the final outputs from STAPRO

==> For the rafter the value of out_b_t_ratio should be < 8.5

==> The value of data item out_b_t_ratio for the rafter conforms to the code
==> According to the code:

the value of out_b_t_ratio should be < 8.5

the actual value of out_b_t_ratio is 2.25

==> For the rafter the value of lm should be =< 1670.91

==> The value of data item lm for the rafter conforms to the code
==> According to the code:

the value of lm should be =< 1670.91

the actual value of lm is 300

==> For the column the value of out_b_t_ratio should be < 8.5

==> The value of data item out_b_t_ratio for the column conforms to the code
==> According to the code:

the value of out_b_t_ratio should be < 8.5

the actual value of out_b_t_ratio is 2.25

==> For the column the value of lm should be =< 1676.27

==> The value of data item lm for the column conforms to the code
==> According to the code:

the value of lm should be =< 1676.27

the actual value of lm is 350

yes

%%% Following are a few interrogation facilities of STAPRO

| ?- **show_applicable_clauses.**

==> 1. 5.3.2

==> 2. 5.3.3

==> 3. 5.3.4

==> 4. 5.3.5

==> 5. 5.3.6

==> 6. 5.3.7

yes
| ?- **advice(section('5.3.2')).**

****
As per Section 5.3.2 :
Plastic design may be used when loading is predominantly

```
static, and, therefore, fatigue is not a design criterion
**********************************************************


yes
| ?- explain(py).


==>     py is design strength



==>     py is needed to calculate the data item lm(_329779)
       as per Section 5.3.5


==>     py is needed to calculate the data item out_b_t_ratio(_329779)
       as per Section 5.3.4

yes
| ?- explain(out_b_t_ratio).


==>     out_b_t_ratio is ratio of b (half the flange width)
        and T (flange thickness) for the outstand element of compression flange



yes
| ?- ^D
Prolog terminated
```

V.2 A sample run illustrating backtracking facility of STAPRO whenever an unconformity to a code pf practice provision occurs

```
Edinburgh Prolog, version 1.5.01 (14 August 1987)
AI Applications Institute, University of Edinburgh

| ?- start,run.


What is the problem?
        1   design
        2   design_check
Type in a number followed by a full stop:2.



What type of design do you want to carry out?
        1   elastic
        2   plastic
Type in a number followed by a full stop:2.
```

Which of the following do you want to carry out ?
1 conformance
2 determination
Type in a number followed by a full stop:**1.**

What type of conformance checking is required ?
1 data item
2 provision
3 general conformance
Type in a number followed by a full stop:**1.**

Which data item is involved ? (The data-item may be one from the list [lm,lt,out_b_t_ratio,in_b_t_ratio])

lm.

What is the value of the data item (in mm.) ?

1700.

==> Checking the applicability criteria of the concerned clause(s)

==> Checking the applicability criteria of clause 5.3.5

==> No applicability criteria for clause 5.3.5 found.

==> The following queries relate to the parameters of the object in question
==> Please input the value of alo: **1500.**
==> Please input the value of dw: **70.**
==> Please input the value of tf: **10.**
==> Please input the value of tw: **10.**
==> Please input the value of wf: **45.**

==> The following queries relate to the parameters of the object in question for the determination of lm
==> Value of rry already present
==> Value of fc already present
==> Value of py already present
==> Value of x already present

==> The value of lm should be =< 1676.27

==> The value of data item lm for the element does not conform to the code

==> According to the code:

the value of lm should be =< 1676.27

the actual value of lm is 1700

Do you want to get suggestions on how to fix the unconformities ?
1 yes
2 no
Type in a number followed by a full stop:**1.**

%%% Backtracking starts here

==> The unconformity of element lm can be fixed by manipulating one or more of the following:

==> 1. af

==> 2. alo

==> 3. aro

==> 4. aw

==> 5. dw

==> 6. fc

==> 7. iry

==> 8. plate_thickness

==> 9. py

==> 10. rry

==> 11. shd

==> 12. steel_grade

==> 13. tc

==> 14. tf

==> 15. tw

==> 16. wf

==> 17. x

Do you want to change wf ?

no.

Do you want to change tw ?

yes.

==> The following queries relate to the parameters of
the object in question
==> Please input the value of tw: **28.**

==> The value of lm should be =< 1700.27

==> The value of data item lm for the element conforms to the code
==> According to the code:

the value of lm should be =< 1700.27

the actual value of lm is 1700

yes
| ?- ^D
Prolog terminated

V.3 A sample run of a part of DETDEX that advises on general considerations in detailed design

Edinburgh Prolog version 1.5.04 (12 September 1988)
AI Applications Institute, University of Edinburgh

| ?- **start,run.**

What is the vertical loading position ?
1 all_over
Type in a number followed by a full stop:**1.**

Is ridge haunch present ?
1 present
2 absent
Type in a number followed by a full stop:**2.**

What is the horizontal loading type ?
1 udl
2 concentrated
Type in a number followed by a full stop:**2.**

What is the horizontal loading position ?
1 eaves_joint
2 all_over

Type in a number followed by a full stop:**1.**

Is eaves haunch present ?
    1 present
    2 absent
Type in a number followed by a full stop:**2.**

Is the section uniform ?
    1 yes
    2 no
Type in a number followed by a full stop:**1.**

Which member is involved?
    1 stanchion
    2 rafter
    3 lateral_restraints
    4 purlins
Type in a number followed by a full stop:**2.**

Is there any secondary member attached to the rafters?
    1 yes
    2 no
Type in a number followed by a full stop:**1.**

What is attached to the rafters ?

purlins.

What is the section-type?
    1 rolled_ub_section
    2 rolled_uc_section
    3 built_up_section
Type in a number followed by a full stop:**1.**

Which is the design criterion concerned?
    1 stability
    2 strength
Type in a number followed by a full stop:**1.**

What is the vertical loading type ?
    1 udl

2 concentrated
Type in a number followed by a full stop:**1.**

Please input position of hinge(first)

x co-ordinate: **12.**

y co-ordinate: **12.**

Please input position of eaves

x co-ordinate: **0.**

y co-ordinate: **4.**

Please input length of rafter : **12.**

yes
| ?- **show_position(rafter,(compression,flange)).**

1. 3/4*rafter_length,upper,flange

2. 1/4*rafter_length,lower,flange

yes
| ?- **show_position(rafter,(tension,flange)).**

1. 3/4*rafter_length,lower,flange

2. 1/4*rafter_length,upper,flange

yes
| ?- **show_gen_requirements.**

==> The rafter stability requirements in this case are:-

1. (position,hinge(no)),checks(elastic)

2. (position,hinge(first)),spacing(lateral_restraint,critical)

3. (position,eaves),spacing(lateral_restraint,critical)

4. (position,all_over),location(lateral_restraint,(compression,flange))

5. rafter_end_eaves_haunch,moment(0.87*mp)

6. haunched_portions,elastic

7. (lateral_restraint,location),compression,flange

8. (lateral_restraint,spacing),eaves,critical

9. (lateral_restraint,spacing),first_hinge,critical

yes
| ?- **show_req(eaves).**

The requirement to be satisfied at eaves is spacing(lateral_restraint,critical)

The requirement to be satisfied at eaves is spacing(lateral_restraint,critical)

yes
| ?- **show_req([0,4]).**

The requirement to be satisfied at [0,4] is spacing(lateral_restraint,critical)

The requirement to be satisfied at [0,4] is spacing(lateral_restraint,critical)

yes | ?- **show_req([12,12]).**

The requirement to be satisfied at [12,12] is spacing(lateral_restraint,critical)

The requirement to be satisfied at [12,12] is spacing(lateral_restraint,critical)

yes
| ?-
Prolog terminated

Appendix VI

This appendix contains listings of runs of the DESCON module as a standalone prototype. The runs are quite small but are illustrative of the types of solutions DESCON may generate using the dependencies between design entities discussed in Chapter 7.

The user inputs and comments are in bold typeface.

Edinburgh Prolog, version 1.5.01 (14 August 1987)
AI Applications Institute, University of Edinburgh

| ?- **start,run.**

What is the problem?
    1   physical_infeasibility
    2   design
    3   design_check
Type in a number followed by a full stop:**1.**

Which of the following is involved?
    1   stanchion
    2   rafter
    3   lateral_restraints
    4   purlins
Type in a number followed by a full stop:**3.**

Which member was restrained by the restraints in question?
    1   stanchion
    2   rafter
Type in a number followed by a full stop:**1.**

What is the section-type?
    1   rolled_ub_section
    2   rolled_uc_section
    3   built_up_section
Type in a number followed by a full stop:**1.**

==> One solution is stanchion section_size(increased)

%%% The following solution is a measure of how *dumb* DESCON is!
%%% Since lateral restraints were removed, DESCON is suggesting to

%%% put them back to solve the problems arising due to their removal!

==> One solution is lateral_restraints

==> One solution is fixed_base

yes
| ?- **start,run.**

What is the problem?
1 physical_infeasibility
2 design
3 design_check
Type in a number followed by a full stop:**1.**

Which of the following is involved?
1 stanchion
2 rafter
3 lateral_restraints
4 purlins
Type in a number followed by a full stop:**4.**

Which member was restrained by the restraints in question?
1 stanchion
2 rafter
Type in a number followed by a full stop:**2.**

What is the section-type?
1 rolled_ub_section
2 rolled_uc_section
3 built_up_section
Type in a number followed by a full stop:**1.**

==> One solution is rafter section_size(increased)

==> One solution is lateral_restraints

==> One solution is fixed_base

%%% Due to insufficient knowledge in the knowledge-base the following
%%% repercussions could not be examined

==> The following repercussions still need to be examined
due to the removal of purlins:

```
        roof_support removed

yes
| ?- ^D
Prolog terminated
```

Appendix VII

This appendix contains a listing of the three modules of INDEX, viz. ALTSEL, DETDEX and DESCON running together. This listing is essentially the three separate runs of these modules given in Appendices III, V and VI put together. However, the integrated nature of INDEX becomes clear by the following listing. The use of DESCON in an integrated system for structural design whenever a change of specification or violation of constraints occurs is also demonstrated by this listing. It should be pointed out that all the provisions of code have not been included. The sole purpose of this appendix is to illustrate the three modules running together. It should also be pointed out that the system (all the three prototypes together) in its present state takes almost twenty minutes to load and the following run took more than half an hour on a Sun 4/110 workstation.

%%% User inputs and comments are in bold typeface

Edinburgh Prolog, version 1.5.01 (14 August 1987)
AI Applications Institute, University of Edinburgh

| ?- **start,run.**

What is the average vertical load (in kN/sq.m)? (The vertical load is assumed to be uniformly distributed over the whole span)

1.18.

What is the eaves height (in metres)?

7.6.

Is there any horizontal load acting on the frame (The horizontal load is assumed to be a concentrated load acting at the knee joints only) ?

1 yes
2 no

Type in a number followed by a full stop:**1.**

Which knee is the horizontal load acting at ?

1 1
2 2
3 both

Type in a number followed by a full stop:**1.**

What is the magnitude of the horizontal load (in kN) ?

15.

What is the purlin spacing (in metres)?

1.25.

What is the span of the building ? (in metres)

25.

What is the pitch allowed for the portal frame (if possible) ? (in radians)

0.3.

Are internal columns allowed in the building ?
1 yes
2 no
Type in a number followed by a full stop:**2.**

What is the type of bases ?
1 fixed
2 pinned
Type in a number followed by a full stop:**2.**

Are cranes to be present in the building ?
1 yes
2 no
Type in a number followed by a full stop:**2.**

****
Trying to find which are the possible lateral load systems :-
****

single-span portal lateral load system is possible

tied portal lateral load system is possible

The frame spacing for the span under consideration should be in the range of 4 to 6 metres and angles or cold formed purlins should be used.
Type in a desired value followed by a full-stop. **4.6**

****

The following alternatives can be considered for
the sides :-
****

One alternative is to just have side rails attached to
the side stanchions.

Another alternative could be have side bracings between
the stanchions.

****
The following alternatives can be considered for the
side cladding :-
****

One alternative for side cladding is to have plastic
coated sheeting all over.

Another alternative for the side cladding is to have
brickwork in one of the following ways :-
1.supported from the structure both vertically and
horizontally,
2.supported only horizontally,
3.self_supporting both vertically and horizontally,
4.self_supporting and also supporting some elements
such as the ends of purlins at the gables.

Another alternative for the side cladding is to have
precast or cast-in-situ concrete wall all over.

****
The following alternatives can be considered for
the roof system :-
****

One alternative for the roofing system is to have
cladding simply over purlins.

Another alternative for the roofing system is to have

bracings between the rafters of the supporting frame.

****
Following are the approximate section sizes for the
different alternatives of feasible lateral load systems :-
****

Following are the feasible sections
for the single span portal alternative :-

381x152UB@52
Zp provided = 959.0

The following are the feasible sections for the tied
portal rafters and columns based on aproximate analysis :-

254x146UB@31
Zp provided = 394.8

Following is the dimensions for the tie based
on approximate analysis :-

60x60x10 angle or a rod of 36mm. dia.

****
The following design constraints should be considered
in the detailed design stage :-
****

The following things should be considered in the
detailed design stage of single span portal alternative :-
1.pitch should be kept low because greater slope
will give rise to greater spread at knees which can cause problems with
cladding,
2.horizontal thrusts should be carefully examined
and the foundation designed accordingly,
3.haunch should be provided at
the eaves and the ridge should be deepened because the maximium bending
moment will occur at the knees.

%%% The detailed analysis of the single span portal frame starts here

****
Starting analysis - preparing data
****

****
Starting the actual analysis
****
Mon May 15 19:41:03 BST 1989

%%% The output of the FORTRAN analysis program is read by PROLOG

ana.out consulted: 4172 bytes 0.40 seconds

%%% The DETDEX module undertakes the standards checking

****
Checking the applicable provisions of the code of practice
****

Is the following a limit state ?

portal_frame ,(plastic,,(sway_stability,whole))

yes.

Is the following a limit state ?

portal_frame(column) ,(plastic,,(stability,whole))

no.

Is the following a limit state ?

portal_frame(rafter) ,(plastic,,(stability,whole))

no.

Is the following a limit state ?

portal_frame(rafter) ,(plastic,,(snapthrough_stability,whole))

no.

==> Checking the applicability criteria of the concerned clauses

==> Checking the applicability criteria of clause 5.3.2
==> Load predominantly static :true or false **true.**

==> Checking the applicability criteria of clause 5.3.3
==> Steel complies to BS4360 :true or false **true.**

==> Checking the applicability criteria of clause 5.3.4

==> No applicability criteria for clause 5.3.4 found.

==> Checking the applicability criteria of clause 5.3.5

==> No applicability criteria for clause 5.3.5 found.

==> Checking the applicability criteria of clause 5.3.6
==> No hinge stiffener located at more than half the depth of the section on either side :true or false **true.**

==> Checking the applicability criteria of clause 5.3.7
==> Any holes on either side of the hinge upto a length equal to the depth of any member either punched or drilled 2mm in diameter undersize and reamed :true or false **true.**

==> Checking the applicability criteria of clause 5.5.3.2

==> No applicability criteria for clause 5.5.3.2 found.

==> Please input the actual value of lm for the column: 565.

==> Please input the actual value of lm for the rafter: 675.

%%% Comparing this run to the run of STAPRO as a standalone system, no user
%%% input is required in this case because all the required inputs are
%%% coming from the other modules

==> The following queries relate to the parameters of column^
for the determination of lm^
==> Value of rry already present^
==> Value of fc already present^
==> Value of py already present^
==> Value of x already present
==> The following queries relate to the parameters of column^
for the determination of out_b_t_ratio
==> Value of py already present

==> The following queries relate to the parameters of rafter
for the determination of lm
==> Value of rry already present
==> Value of fc already present
==> Value of py already present
==> Value of x already present

==> The following queries relate to the parameters of rafter
for the determination of out_b_t_ratio
==> Value of py already present

%%% Following are the outputs from the standards checking by STAPRO

==> For the portal_frame the value of 2-1-node_displacement should be =< 0.0076

==> For the portal_frame the value of 4-1-node_displacement should be =< 0.0076

==> For the column the value of lm should be =< 1169.9

==> For the rafter the value of lm should be =< 1169.9

==> For the column the value of out_b_t_ratio should be < 8.5

==> For the rafter the value of out_b_t_ratio should be < 8.5

%%% Interrupt

| ?- **change_spec.**

What type of parameter do you want to change ?
1 primary
2 secondary
Type in a number followed by a full stop: **2.**

%%% The program carries on with the things unaffected by the change

==> The value of data item 2-1-node_displacement for the portal_frame does not conform to the code
==> According to the code:

the value of 2-1-node_displacement should be =< 0.0076

the actual value of 2-1-node_displacement is 8.11372e-03

%%% DESCON carries on with the change of specification problem

Which of the following describes the new constraint most closely ?
1 physical_infeasibility of some existing member/sec_member
2 additional_cons
Type in a number followed by a full stop:**1.**

Which of the following categories does the involved object belong to ?
    1 member
    2 sec_member
Type in a number followed by a full stop:**2.**

Which of the following is involved ?
    1 lateral_restraints
    2 purlins
Type in a number followed by a full stop:**1.**

%%% The program carries on with the things unaffected by the change

==> The value of data item 4-1-node_displacement for the portal_frame conforms to the code ==>
According to the code:

the value of 4-1-node_displacement should be =< 0.0076

    the actual value of 4-1-node_displacement is 5.55500e-03

==> The value of data item lm for the column conforms to the code
==> According to the code:

the value of lm should be =< 1169.9

    the actual value of lm is 565

==> The value of data item lm for the rafter conforms to the code
==> According to the code:

the value of lm should be =< 1169.9

    the actual value of lm is 675

==> The value of data item out_b_t_ratio for the column conforms to the code
==> According to the code:

the value of out_b_t_ratio should be < 8.5

    the actual value of out_b_t_ratio is 6.14

%%% DESCON proceeds with the solution of change of specification problem

Which member was restrained by the restraints in question?
    1 column
    2 rafter
Type in a number followed by a full stop:**1.**

%%% DESCON generates solutions based on heuristics

==> To overcome the effects of excessive 2-1-node_displacement
one solution is to introduce ridge haunches

==> To overcome the effects of decreased stability
due to the removal of lateral_restraints one solution is to have bases(fixed)

%%% DESCON generates solutions based on dependencies

==> To overcome the effects of decreased stability due to the removal of lateral_restraints one solution is column aro(increased)

==> Please input another section with a higher
section modulus: **1046.**

%%% DESCON passes back the control to DETDEX and picks up the new section

Following are the feasible sections
for the single span portal alternative :-

406x178UB@54
Zp provided = 1046

%%% DETDEX passes the new section for re-analysis

****
Starting analysis - preparing data
****

****
Starting the actual analysis
****
Mon May 15 20:04:59 BST 1989

%%% The new output of FORTRAN read into PROLOG

ana.out consulted: 2176 bytes 0.40 seconds

%%% DETDEX checks the new section

==> For the column the value of lm should be =< 1403.15

==> For the rafter the value of lm should be =< 1403.15

==> The value of data item lm for the column conforms to the code
==> According to the code:

```
 the value of lm should be =< 1403.15

          the actual value of lm is 565


==>  The value of data item lm for the rafter conforms to the code
==>  According to the code:

 the value of lm should be =< 1403.15

          the actual value of lm is 675

yes
| ?- ^D
Prolog terminated
```

Appendix VIII

This appendix contains a listing of a run of a FORTRAN program written in the earlier part of the work for the design of a simply supported, unstiffened, non-hybrid, welded plate girder. The program stores the results of each run in three separate databases. Based on the results stored in these databases, the program is able to advise the user whether his assumption of the span-to-depth ratio is likely to cause failure due to bending or shear. The three databases are meant to store different values as described below:

1. the unsafe dimensions against bending along with the maximum bending moment and shear force;
2. the unsafe dimensions against shear along with the maximum bending moment and shear force; and
3. the safe dimensions along with the maximum bending moment and shear force.

The first two databases assist the user in making assumptions as regards the span-to-depth ratio. This is done by matching the assumed value of the span-to-depth ratio for the corresponding values of maximum bending moment and shear force in the respective databases. The third database helps in picking up the safe dimensions for a particular problem by matching the maximum bending moment and shear force, thus, saving time and effort. The accompanying sample run illustrates the feedback from the past designs quite clearly.

```
Command:RUN TEST:
PLEASE USE UPPER-CASE LETTERS ONLY
DO YOU WANT TO DESIGN A PLATE GIRDER?
Data:YES
WHAT IS THE SPAN(in metres)?
Data:18.0-
WHAT IS THE TYPE OF IMPOSED LOADING ?
TYPE 'UDL' FOR UNIFORMLY DIST. LOAD,
TYPE 'CLS' FOR CONC. LOADS ,
TYPE'UDL+CLS'FOR A COMBINATION OF UNIFORMLY DISTRIBUTED
LOAD AND CONCENTRATED LOADS.
Data:CLS
HOW MANY CONCENTRATED LOADS ARE THERE?
Data:3
WHAT IS THE MAGNITUDE OF THE CONCENTRATED LOADS(in kNs.)?
Data:700,700,700
WHAT ARE THE DISTANCES OF THE LOADS FROM
THE LEFT END(in metres) ?
Data:4.5,9.0,13.5
WHAT IS THE DESIRED THICKNESS OF THE PLATES TO
BE USED(in mm.)?
Data:40.0
REACTION1=  1050.000    REACTION2=  1050.000
MAX. S.F.=  1050.000
MAX. B.M.=  6300.000
HOW DEEP DO YOU WISH THE GIRDER TO BE(in metres)?
IT SHOULD NORMALLY VARY BETWEEN 1/8 AND 1/12 OF THE SPAN
IN YOUR CASE,1/8 OF SPAN=2.250 1/12 OF SPAN=1.500
Data:2.25
*************************************************
FOLLOWING ARE THE REVISED REACTIONS,MAX. S.F.,
AND MAX. BENDING MOMENT INCLUDING THE SELF-WT.
OF THE GIRDER
*************************************************
REACTION1=  1071.967    REACTION2=  1071.967
MAX. S.F.=  1071.967
MAX. B.M.=  6396.848
HOW WIDE DO YOU WISH THE FLANGE TO BE(in metres)?
SELECT A VALUE BETWEEN WF1 AND WF2.
WF1= 0.4500000      WF2= 0.2250000
Data:0.350
****************************************
       FIRST TRIAL DIMENSIONS(in mm.)
****************************************
DG=  2250.000
WF=  350.0000       DF=  29.99998
TW=  9.999997       DW=  2190.000
*****************************************
THE DIMENSIONS ARE NOT SAFE AGAINST BENDING
WHICH DIMENSION OF THE FLANGE DO YOU WISH TO
TO ALTER ? TYPE IN DEPTH OR WIDTH
Data:DEPTH
```

```
NOW,THE DIMENSIONS ARE SAFE AGAINST BENDING
THE DIMENSIONS ARE NOT SAFE AGAINST SHEAR BUCKLING
WHICH WEB DIMENSIONS DO YOU WISH TO ALTER ? TYPE IN
DEPTH OR THICKNESS
Data:THICKNESS
NOW,THE DIMENSIONS ARE SAFE AGAINST BENDING
NOW,THE DIMENSIONS ARE SAFE AGAINST SHEAR BUCKLING
ALSO
THE SECTION IS SAFE AGAINST DEFLECTION AS WELL
********************************
     REVISED DIMENSIONS (in mm.)
********************************
DG=  2269.999
DF=  39.99998     WF=  350.0063
DW-  2189.999     TW=  19.99998
```